From Lectern to Laboratory

Selected Works by W. Nikola-Lisa

Bein' With You This Way

Can You Top That?

Dear Frank: Babe Ruth, the Red Sox, and the Great War

Dog Eared: A Year's Romp Through the Self-Publishing World

Dragonfly: A Childhood Memoir

Folk Stories

Gaya Lives in a Blue House

How We Are Smart: A Multicultural Approach to the Theory of Multiple Intelligences

Magic in the Margins: A Medieval Tale of Bookmaking

Shake Dem Halloween Bones

Summer Sun Risin'

The Men Who Made the Yankees

Till Year's Good End

From Lectern to Laboratory

How Science and Technology Changed the Face of America's Colleges

W. Nikola-Lisa

Gyroscope Books
Chicago

Published by Gyroscope Books
Copyright © 2019 by W. Nikola-Lisa
All rights reserved.

Gyroscope Books® is the independent publishing arm of NikolaBooks Productions. For information about permission to reproduce selections of this book, email the author at nikolabooks@gmail.com.

For orders by U.S. trade bookstores, retail outlets, and public libraries, please contact the Ingram Content Group: Tel: Retail (800) 937-8000; Libraries (800) 937-5300; or visit www.ingramcontent.com.

All images from the "Magasin Pittoresque" collection at ShutterStock. Edited by Beth Lynne, BZ Hercules Editing and Consulting Services. Designed by Deb Tremper, Six Penny Graphics. PCIP Data prepared by The Donohue Group, Inc.

The quotation from Moravian philosopher, theologian, and pedagogue Johann Amos Comenius that heads the Preface comes from *Papers and Addresses of William Henry Welch, Vol. 3* (Baltimore: The Johns Hopkins University Press, 1820), p. 202.

Summary. In this well-documented account, W. Nikola-Lisa explores how science and technology changed the face of America's colleges during the nineteenth century, bringing much needed reform to the classical liberal arts curriculum while, at the same time, suggesting new approaches to instruction.

Library Congress Control Number: 2019902991

ISBN: 978-0-9972524-8-4 [hardcover]
ISBN: 978-0-9972524-9-1 [paperback]
ISBN: 978-0-578-46071-0 [e-book]

Publisher's Cataloging-In-Publication Data
(Prepared by The Donohue Group, Inc.)

Names: Nikola-Lisa, W., author.
Title: From lectern to laboratory : how science and technology changed the face of America's colleges / W. Nikola-Lisa.
Description: Chicago : Gyroscope Books, [2019] | Includes bibliographical references and index.
Identifiers: ISBN 9780997252484 (hardcover) | ISBN 9780997252491 (paperback) | ISBN 9780578460710 (ebook)
Subjects: LCSH: Universities and colleges--Curricula--United States--History--19th century. | Science--Study and teaching (Higher)--United States--History--19th century. | Technology--Study and teaching (Higher)--United States--History--19th century. | Education--Research--Laboratories--History--19th century.
Classification: LCC LB2361.5 .N55 2019 (print) | LCC LB2361.5 (ebook) | DDC 378.1990973--dc23

To my father—
William Henry Nikola, B.S.
Metallurgy and Minerals Engineering
Colorado School of Mines

And my uncle—
Herbert Charles Nikola, Ph.D.
Soils Engineering
Rutgers

Contents

Preface ix

CHAPTER 1 | *Land of Living Waters* 1

CHAPTER 2 | *Reading, Writing, Refmetic* 25

CHAPTER 3 | *Schools and Schoolmasters* 43

CHAPTER 4 | *The Leaven of Improvement* 61

CHAPTER 5 | *"Old Sheff"* 85

CHAPTER 6 | *An Impenetrable Thicket* 107

CHAPTER 7 | *Bridge to the Future* 129

CHAPTER 8 | *Outshine Them All* 151

CHAPTER 9 | *Objects and Plan* 169

CHAPTER 10 | *The New Education* 187

CHAPTER 11 | *With His Own Hands* 211

Significant Events 229

Chapter Notes 235

Bibliography 267

Name Index 281

Preface

> Men must be instructed in wisdom so far as possible, not from books, but from the heavens, the earth, the oaks and the beeches; that is, they must learn and investigate the things themselves, and not merely the observations and testimonies of other persons concerning the things.
> —Johann Amos Comenius

A cursory look at elementary through high school curriculum materials confirms, without question, that we are in the age of STEM—science, technology, engineering, and mathematics (or, as progressive educators prefer, STEAM, adding arts to the lineup). You'll see the STEM acronym splashed all over student textbooks, scope and sequence charts, laboratory manuals, and even on posters in school hallways. It shouldn't be a surprise: after all, we live in the information age where students not only must be able to think critically, but also must be able to apply concepts from science and technology to solve the complex problems they face today. Yes, the world of STEM is upon us, and educators around the world are turning their attention to it accordingly.

But this is not the first time that we've heard the clarion call for more instruction in science and technology. Think of the impact that the launch of the Russian spaceship Sputnik had on the American scientific—and political—community in the 1950s. But even that momentous, and for some frightening, event is not the starting point of STEM. The call for more instruction in science, technology, engineering, and mathematics has been with us in varying degrees and intensity since the beginning of the industrial revolution. It's just that it has not always been met with open arms. Prior to the middle of the nineteenth century, college presidents and faculty resisted the introduction of courses of study in science, especially if such programs favored the applied sciences, fearing that such practical approaches, bordering on the vocational or trade school, would dilute the classical liberal arts curriculum.

It is this struggle, the struggle to integrate science and technology into the college curriculum of the early- and mid-nineteenth century, that is the focus of this book. However, rather than organize the work around the four disciplines that make up the STEM acronym, I've decided to take another tact: to write a biography, but not in the strict sense of the word—a treatise that explores the details of an individual's life circumstance and the inner and outer forces that conspired to shape his or her destiny. The story I intend to tell, though it revolves around the early educational and work experiences of one individual, is manifold and will take us to many different corners of this struggle, some that will implicate our subject directly, and others that won't. In this sense, then, the following work is more of a cultural biography than a personal biography.

Be that as it may, the individual I have in mind, who will act as the focal point of this study, is Edward Charles Pickering, a man of great accomplishment, who distinguished himself as the fourth director of the Harvard Observatory with the introduction of photography and spectrometry into the rapidly changing world of nineteenth-century astronomy. However, it is not Pickering's work at the Harvard Observatory that interests me per se, as interesting as it is for what he and others, especially women, achieved during his tenure as director of the observatory. For a comprehensive look at the achievements of women working both as "human computers" and astronomers at the Harvard Observatory under Edward Pickering, I recommend Dava Sobel's *The Glass Universe: How the Ladies of the Harvard Observatory Took the Measure of the Stars*.

What I find of interest in Edward Pickering's story are the educational opportunities available to Edward, opportunities of which he and his parents readily took advantage, often through the mentoring of close family friends. But why choose Edward Pickering in the first place? There must be a number of individuals from whom to choose, individuals who could serve the same purpose. True, I'm sure. But there are several reasons why I chose Edward Pickering as the subject of this study: he was born at the right time (in the middle of the nineteenth century), in the right city (Boston), and to parents of means (and of no uncertain pedigree). Equally important, however, is what he demonstrated as a youth: a facility with his hands, a knack for problem solving, and an interest in technology, which he would continue to develop at the Lawrence Scientific School at Harvard and later at the newly-established Massachusetts Institute of Technology.

In other words, to understand Edward Pickering's story is to understand the story of how science and technology struggled to find a place at the table of America's colleges and universities during the nineteenth century. But reader beware, as the son of a wealthy Boston family, Edward Pickering's story is not the story of everyone. It is not the story of the sons and daughters of first-generation immigrant families who arrived on Boston's shores in waves throughout the nineteenth century. It is not the story of the descendants of free blacks living in cramped quarters on the north slope of Boston's Beacon Hill. Nor is it the story of young women, from any rank of society, who lacked the same educational and work opportunities available to our subject. And, yet, given these considerations—or, delimitations, if you will—Edward Pickering's story is compelling as it traces the arc of a growing interest in science and technology that defined much of the nineteenth century.

It is a story that has its roots in some of the nation's oldest colleges—Harvard, Yale, the College of New Jersey (Princeton), William and Mary, King's College (Columbia), and Dartmouth. Although the struggle of science and technology to find its rightful place in higher education involves many of these institutions—and others as well (Jefferson's University of Virginia, for example)—the epicenter of this story is Boston. From the time members of the Massachusetts Bay Colony founded the city in the 1630s, Boston has been the hub of growth and change in a variety of fields—social, political, economic, and educational.

Indeed, Boston has been a testing ground for intellectual, civic, and educational ideas from the early seventeenth century, beginning with the founding of the first public Latin grammar school (Boston Latin, 1635) and the nation's first college (Harvard, 1636). It was also the epicenter of Josiah Holbrook's American Lyceum Movement and Horace Mann's Common School Movement. Nor has it shied away from introducing science and technology into the secondary and post-secondary curriculum: Boston was the site of the nation's first public vocationally-oriented high school (English High School, 1821); Boston also established one of the earliest scientific schools affiliated with a university (Lawrence Scientific School at Harvard, 1847); and the city remains the location of one of the nation's preeminent scientific institutes (Massachusetts Institute of Technology, 1865).

It is my hope, then, by following Edward Pickering's education and early professional experiences in Boston and neighboring Cambridge, that I am able to shed light not only on the forces that contributed to the rise of science and technology programs at American colleges and universities, but also on the essential role that Boston's intellectual and civic communities played in this struggle. This book, however, is in no way a comprehensive look at those forces, nor is it—as I mentioned earlier—a strict biography, personal or cultural: it is merely a window into that world afforded to us through the educational and professional opportunities of Edward Pickering, opportunities unique to him, his family, and to his time.

CHAPTER 1 | *Land of Living Waters*

> *The news that the new capitol building would be constructed on the crest of Beacon Hill immediately produced other changes in the rustic surroundings of nearby Park Street and Tremont Street where hay carts rumbled and cattle still grazed. Within a dozen years the whole area was in the midst of rapid development, transforming the old Puritan town of wood and thatch into a new Federalist capital of brick and granite.[1]*
> —Thomas H. O'Connor

We know very little about the early life of Edward Pickering. No definitive biography exists. What has been written usually begins after Pickering accepted the directorship of the Harvard Observatory, which he did in November of 1876, beginning his official duties at the observatory on the first of February 1877. His work at the observatory, which at the time was one of the most important centers for astronomical research in the world, is well documented both by internal annual reports and external analyses.

What we don't know is how Pickering spent his childhood, other than casual references to his life as the child of an established Boston family, his family's Beacon Hill residence, and Pickering's graduation from the prestigious Boston Latin School at the age of sixteen. In his biographical memoir written for the National Academy of Sciences in 1932, Solon Bailey, an early employee and close associate of Edward Pickering, wrote:

> Mr. Pickering was fortunate in his heritage. Of a family always prominent in New England history, he was heir neither to riches nor to poverty, but to splendid opportunity, which he eagerly grasped. From early youth to old age, his

zeal in the pursuit of scientific problems was unbounded. His education was begun in private schools, but later carried forward at the Boston Latin School. He had small love of the classics and gave them scant attention. In the Lawrence Scientific School, however, he entered upon his work with that enthusiasm which marked all the activities of his mature life. He was graduated from this school summa cum laude at the age of nineteen, and was immediately appointed Instructor of Mathematics in that institution. A year later he became Assistant in Physics at the Massachusetts Institute of Technology, and in the following year Thayer Professor of Physics, a position which he held until he became Director of the Observatory.[2]

This is a typical biographical sketch, which skips over Pickering's childhood, mentioning only briefly his graduation from Boston Latin, and focuses on his years at Lawrence Scientific School, the Massachusetts Institute of Technology, and the Harvard Observatory, with the latter receiving the closest attention. In order to understand Edward Pickering's life in Boston as a youth, we'll need to piece together his childhood from the few scraps of information we have and from practices common to the Boston area during his formative years. Although there is not a lot of information about Pickering and his family life in Boston, Pickering's lineage has received close scrutiny in a three-volume genealogy published in 1897 by Harrison Ellery and Charles Pickering Bowditch, two descendants of the Pickering line.[3] The genealogy spans nine generations and reveals that Edward Charles Pickering descended from one of New England's oldest and most distinguished families.

The family history starts with John Pickering, Edward's first American ancestor, who emigrated from Yorkshire, England with his wife Elizabeth in 1636, and settled in Salem, Massachusetts, after spending several years in Ipswich. Within a few years, John Pickering had acquired several dozen acres of land outside of Salem, on which he established the family home.[4] Two of John Pickering's children are known to have survived childhood: John and Jonathan. After their father died, John, the oldest son, inherited most of the family property, residing on the farmstead with his wife Alice and five children—John, Benjamin, William, Elizabeth, and Hannah. Not only was he a capable farmer, continually adding acreage to the family estate, but he was also a devoted civic servant, filling a variety of offices, including constable and town councilman. He also served as a lieutenant in the local militia, distinguishing himself in the Indian War of 1675, also known as King Phillip's War. Like his father, John Pickering bequeathed the family farm to his oldest son, also named John, who became a well-respected citizen of Salem, filling the office of city councilman as well as representing the area in the Massachusetts General Court. He too expanded the family estate, willing portions of it upon his death to his wife Sarah and their six children—John, Theophilus, Timothy, Lois, Sarah, and Eunice.

At this point, Edward Pickering's family line descends not from John and Sarah's oldest son, John, or their second eldest, Theophilus, but from their third child, Timothy, who was born in Salem in 1702. Known locally as "Deacon Pickering" for his involvement in Salem's Trinitarian Church, Deacon Pickering was a man of firm conviction, moral character, and great piety. He was also industrious, enlarging the Pickering family estate, which he had inherited from his father, and through his frugality, providing education to his nine children, including sending his two sons—John and Timothy—to Harvard College, the first in his family to do so.

Perhaps the most illustrious member of Edward Pickering's ancestral line is his great-grandfather Col. Timothy Pickering, Deacon Pickering's youngest son. As the youngest son, Col. Pickering inherited a much smaller share of the family estate, most of it going to his oldest brother, another John Pickering. As such, rather than spend his life on the family farm or in business affairs of his own, Timothy joined the Fourth Military Company of Salem, rising to the level of captain in 1769. For the next few years, Captain Pickering served Salem and the surrounding county in several capacities: as city councilman, town clerk, justice of the peace, and register of deeds.

Never straying too far from his interest in military affairs, Timothy was appointed Colonel of the First Regiment of Essex County Militia in 1775, and, due to his law background, Adjutant-General of the United States Army two years later. A frequent visitor to General Washington's headquarters, Col. Pickering soon became a favorite of the Continental Army's high command. The association led to his appointment as one of three members of the Continental Board of War, and then to Quartermaster-General of the Army. Although he retained the rank of colonel, he was paid at the higher rate of brigadier-general. After the war, Col. Pickering found himself once again in the employment of the federal government: this time as Postmaster-General, a position he accepted in 1791. Four years later, he was appointed Secretary of War, which included oversight over the Army, the Navy, and Indian Affairs. During the same year, Col. Pickering became acting Secretary of State under President Washington.

After serving nine years in two different administrations—under Presidents Washington and Adams—Col. Pickering returned to private life. But that did not last long. In 1803, a year after he lost a bitter fight to represent the Federalist Party in the U.S. House of Representatives, the Massachusetts General Court elected Col. Pickering to succeed retiring senator Dwight Foster. He was re-elected by popular vote two years later, whereupon he began a six-year term in the U.S. Senate. After his tenure in the Senate, Col. Pickering was elected by an overwhelming majority to represent Essex County North District in the Thirteenth U.S. Congress. Col. Pickering retired from public service in 1818 and for the next decade, before his death in 1829, lived in Salem, visiting the family farm in Wenham quite often. Like his ancestors before him, farming was in his blood, and living close to the land at the end of his life pleased him more than anything else. Col. Pickering married Rebecca White of Bristol, England, and raised seven children to adulthood—John, Timothy, Henry, Charles, William, George, and Octavius.

Of the seven children, Col. Pickering's eldest son, John, Edward Pickering's grandfather, concerns us the most. Not only does he represent the direct line of descent to Edward, but also he is the most fascinating, accomplished, and esteemed of his siblings. In many ways, he is just like his father—industrious and ambitious. Born in Salem in 1777, John Pickering grew up under the watchful eye of his unmarried uncle, who occupied the family residence in Wenham north of Salem. After a childhood of public schooling and private tutors, John left for Cambridge, where he studied law at Harvard. After graduating from Harvard in 1796, John moved to Philadelphia to practice law in the office of Edward Tilgman. But eight months later, he found himself on a boat headed to Lisbon, Portugal, to fulfill the duties of Secretary of Legation. Two years later, he moved to London, where he became Secretary of Legation under Rufus King, the U.S. Minister to England.

John Pickering returned to Salem at the end of the summer in 1801, working in the law office of his cousin, Samuel Putnam. After being admitted to the Massachusetts Bar in 1804, he left his cousin's practice and opened his own office in Salem. But he never forgot his experience abroad, especially his growing interest in language, of which he had mastered not only Latin and Greek, but also a variety of other languages, including Hebrew. His intimate knowledge of language, which included the languages of many North American Indians, drew the attention of several Harvard professors, so much so that, between 1806 and 1812, he was twice offered a professorship to teach language at his alma mater.[5] Pickering declined both offers, not only because of his successful law practice, but also because of his expanding engagement in politics, which included representing Salem in the Massachusetts General Court and Essex County in the Massachusetts State Senate. In 1818, he was appointed to the prestigious Governor's Council. Due to his increasing legislative responsibilities, he moved to Boston at the end of 1826 and soon was serving as city alderman, city solicitor, and state senator representing Suffolk County.

In 1832, John Pickering moved to 73 Beacon Street with his wife Sarah White Pickering, where he resided until his death. Like his father, John Pickering was a learned and accomplished scholar, a man of letters, and the author of several books, including a critically-acclaimed lexicon of the Greek language and the first dictionary of American idioms. Along with Harvard (where he served as a member of the Board of Overseers), he had affiliations with the Boston Latin School, the Boston Society for the Diffusion of Useful Knowledge, the Massachusetts Society of the Cincinnati, and the American Academy of Arts and Sciences. John Pickering died on May 5, 1846, two-and-a-half months before his grandson, Edward Charles Pickering, was born. He was survived by his wife Sarah and three adult children—Charles, Sarah, and Edward.

It is the youngest sibling, Edward Pickering, who concerns us the most, for he is Edward Charles Pickering's father. Like many of his male forebears, Edward Pickering—John Pickering's youngest son—attended Harvard, graduating in 1824 at the age of 17. After working in New York City for several years, Edward returned to Boston to study law with his uncle. Rather than practice law, however, Edward developed an interest in financial management and wound up working for several large organizations, including the Boston and Maine Railroad, the United States Hotel Company, and the Boston Society of Natural History. But it is not the various business and civic organizations that the elder Edward Pickering worked for that tell the story of his climb up Boston's social and economic ladder. Rather, it is Edward's successive residences in and around Boston's financial district that illustrate that ascent, a record of which is contained in *The Boston Directory*, an annual directory of the addresses of every Boston resident and business, first published in 1789.[6]

The first street Edward lived on after returning to Boston from New York was Franklin Street, which is also the same street on which Edward's uncle, John Pickering, lived. We know this from John Pickering's daughter's memoir of her father, where she writes: "On removing to Boston a house was taken in Franklin Street (No. 5), on the southern side, between Hawley Street and Washington Street, in a new brick block of four houses, owned by Mr. Barnabas Hedge, of Plymouth."[7] Since there is no house number listed for Edward Pickering until the

1830 directory, we can only assume that he stayed with his uncle on Franklin Street when he moved to Boston.

Franklin Street is about two blocks east of Tremont Street, which runs along the eastern edge of Boston Common. It is also close to the Old Granary Burying Ground and the Boston Athenaeum. Located between Beacon Hill's upscale South Slope neighborhood and Boston's thriving downtown business district, a house on Franklin Street was more for convenience than for advertising one's social and economic status. Edward's next address, listed in the directory's 1831 edition as 18 Pearl Street, appears to be another residence of convenience. It is the address of Octavius Pickering, John Pickering's youngest brother and Edward's uncle. Edward lived at 18 Pearl Street with his uncle for five years until Octavius moved to 70 Beacon Street in 1837, just down the street from his brother's residence at 73 Beacon Street. By establishing their family homes in Beacon Hill across from Boston Common, John and Octavius Pickering epitomized the social and economic ladder the Pickering family was determined to climb.

Edward, too, was determined to climb that ladder, and, like his father and his uncle, he had the wherewithal to do so: he was a graduate of Harvard's law school and a member of a prominent New England family. For Edward, the next step up that ladder is reflected in his next address: 43 Bowdoin Street. The year is 1842, five years after Octavius relocated to Beacon Street. It is a significant year for the elder Edward for several reasons. First of all, Bowdoin Street, a mere stone's throw from the Massachusetts State House, is closer to the affluent neighborhoods of Beacon Hill's South Slope district, a sure sign of Edward's aspiring upward mobility.

The move to Bowdoin Street also marks the beginning of a new chapter in Edward's life. It is the year of his marriage to Charlotte Hammond, the daughter of Daniel Hammond, a successful Boston merchant. Later that year, the couple would celebrate the birth of their first child, Ellen Hammond Pickering. Edward and Charlotte lived at 43 Bowdoin Street for the next seven years. During this time, they witnessed the birth of a second child, Edward Charles Pickering. They would have a third child, William Henry Pickering, but he would not be born while the family lived at 43 Bowdoin Street. By William's birth, some twelve years later, Edward and Charlotte Pickering lived at a different address—74 Mount Vernon Street—in Beacon Hill's tony South Slope neighborhood. They would reside at this address until their children were grown. It was an address of which they were extremely proud, as it reflected their ultimate goal: acceptance into the highest circle of Boston's elite families.

BEACON HILL FIGURES PROMINENTLY in the early history of Boston, for it was on the southern slope of Beacon Hill that the earliest European colonist, William Blackstone—or Blaxton—settled. He was a reclusive Episcopal clergyman from England who came to the New World in 1623 as the chaplain of a group of colonists led by Robert Gorges that settled on the southern shore of Massachusetts Bay. After the settlement failed, Blackstone wandered northward, settling on the western side of the Shawmut Peninsula around 1625, not far from the fresh water springs for which the peninsula was known.[8]

Blackstone's quiet, pastoral life lasted only five years. In 1630, he invited a small group of English settlers led by John Winthrop to join him on the peninsula. Winthrop and his followers had settled across the Charles River, but their fresh water supply was not reliable. After an epidemic caused by fetid water sickened many of Winthrop's followers, Winthrop and several other colonists visited Blackstone to inquire about Shawmut Peninsula's potable water supply. When Blackstone described the fresh water source as clean and spring-fed, Winthrop and his followers packed up and moved across the Charles River. As Winthrop's settlement thrived, Blackstone's quiet, reflective life withered, so much so that, in 1634, he packed up and moved to the woods of Rhode Island.

Over the next one hundred and fifty years, Boston, named after a town in Lincolnshire, England, from which many of the first English immigrants hailed, developed along two prominent axes: Orange Road, which ran the length of the peninsula into the city center (via Newbury, Marlborough, and Cornhill), and King Street, which led from Cornhill to the inner harbor, with most of the population settling east of Orange Street facing the harbor. If there were an epicenter of colonial Boston, at least for the ruling class, it would have been at the intersection of King Street and Cornhill. That was where the Old State House was located, on a rise overlooking the busy, half-moon-shaped cove on the seaward side of the peninsula. But expansion was inevitable, and the green light to that expansion was construction of the New State House on pastureland owned by John Hancock, the state's first-elected governor. Both the massive structure and its placement pointed toward Boston's future development—west from Beacon Hill to the Charles River and south toward Back Bay. It is Beacon Hill that concerns us here, especially the land west of the state capitol.

Beacon Hill derived its name from the location of a signal beacon built in 1634 atop the highest point on the Shawmut peninsula. The beacon, however, which at the time was nothing more than a tar-filled iron skillet bracketed to the top of a wooden mast, was not a beacon warning sailors of an approaching coastline; its purpose—explicitly stated by the colonial government—was to warn residents of enemy attack. By the early 1800s, however, the beacon was gone—blown down by gale winds in 1789—and so was most of the hill, leveled to make room for residential expansion. Although no more than a square mile or two, by the middle of the century, Beacon Hill was comprised of two socially distinct neighborhoods, with Pinckney Street serving as the functional north-south dividing line between the North Slope's working-class immigrant community (including, at the time, the largest concentration of African-American families above the Mason-Dixon Line) and the South Slope's well-heeled upper-class families, often referred to as "Boston Brahmins." A third neighborhood, known as Flat of the Hill, extended west of Charles Street to the banks of the Charles River and was home to a variety of tradesmen who lived and worked in the gritty blue-collar community that served Boston's nearby wealthier neighborhoods. Of the three neighborhoods, South Slope concerns us the most as it contained the childhood home of Edward Pickering.

Bordered by Bowdoin Street and the grounds of the Massachusetts State House to the east, Pinckney Street to the north, Charles Street to the west, and Beacon Street to the south, Beacon Hill's South Slope neighborhood attracted some of Boston's earliest settlers, among them John Singleton Copley, one of Boston's earliest inhabitants, as well as Thomas Hancock, John Hancock's uncle whose widow bequeathed their estate, known as Hancock Manor, to her nephew. Christian Remick captured the mood of Beacon Hill's pastoral nature in a 1768 watercolor titled "A Prospective [sic] View of Part of the Commons."[9] The sketch depicts several structures owned by Copley, a successful portrait painter himself, as well as Hancock Manor overlooking an expanse of pastureland in which a wide array of residents co-mingle—picnickers, horseback riders, Sunday strollers, even a smart-looking contingent of British soldiers. With the completion of the New State House in 1798, South Slope began to attract wealthy families who enjoyed the bucolic views of the Charles River, the neighborhood's sunny southern exposure, and its proximity to Boston's burgeoning financial district.

The attraction was fueled by the efforts of the Mount Vernon Proprietors, a highly charged group of land developers led by Jonathan Mason and Harrison Gray Otis. From the outset, the Mount Vernon Proprietors embarked on an ambitious plan that involved developing eighteen-and-a-half acres of land purchased west of the New State House. To prepare the area for residential development, the group proposed sheering off the tops of Mount Vernon and Beacon Hill, using a combination of manual labor and mechanical innovation in order to transport the remnants of the prominences to several low-lying wetlands, including Mill Pond to the north and the Charles River to the west. By the time Edward Pickering's parents moved into their new residence on Mount Vernon Street, Beacon Hill's South Slope neighborhood had become the indisputable seat of wealth and power, home to Boston's upper crust blue-chip families, an area known for its Federal-style mansions and Greek-Revival townhouses.[10]

LET'S IMAGINE THAT THE year is 1856, the year that Edward Pickering turned ten years old and was preparing to enter Boston Latin. He would be living with his family at 74 Mount Vernon Street, having moved there in 1849 from the family residence on Bowdoin Street behind the Massachusetts State House. If we were to take a walk around Edward's neighborhood in 1856, what would we see? How would it appear over one hundred and fifty years ago? We begin our tour on the front steps of 74 Mount Vernon Street, Edward Pickering's childhood home. The first things we notice are two townhouses—Numbers 87 and 89—across the street. Charles Bulfinch bought the lots from Harrison Gray Otis, who lived next door at 85 Mount Vernon in a Federal-style mansion that Bulfinch designed for him. Like Otis, Mason, and other Mount Vernon Proprietors, Bulfinch intended to do a little speculating of his own: he'd buy two lots, put up two townhouses, live in one of them, and sell the other. It didn't quite turn out that way, however. Finding himself in financial trouble, Bulfinch sold the properties before they were completed.

From here, we turn left and head west along Mount Vernon Street in our circuit around Edward Pickering's South Slope neighborhood. We don't have to walk very far before we come to one of the residential gems of the city—Louisburg Square. Although called a "square," a term that usually denotes a public space, Louisburg Square is a private area of exclusive residences fronting a rectangular green space between Mount Vernon and Pinckney Streets. Named after the 1745 military campaign led by Massachusetts' militiaman William Pepperrell that destroyed the French-Canadian fortress of Louisburg on what is present-day Cape Breton Island, Louisburg Square was designed as a model of townhouse development in the mid-1820s, with most of the Greek-Revival townhouses lining either side of the tree-lined square completed between

1834 and 1844. Two lines of shoulder-to-shoulder, redbrick townhouses face a fence-enclosed oval that bursts with greenery, thanks to the stewardship of the Louisburg Square Proprietors, which claims the distinction of being the nation's first homeowners association.

Continuing past Louisburg Square, we take a left at Cedar Street. Like most of the streets in Beacon Hill's South Slope neighborhood, Cedar Street is delightfully picturesque with its brick townhouses nestled together. Despite the density of the housing units, the architectural details still grab your eye: wooden shutters frame mullioned windows, iron grillwork outlines faux balconies above entryways bordered by shallow keystones and wooden pilasters, flowerboxes cantilever from first-floor windows along tree-lined brick sidewalks, and of course, there's the occasional gas-lit streetlamp taken right out of a Dickens' novel. Halfway down the block, we stop at another attraction—Acorn Street. Although the narrow lane doesn't add much to the menu of architectural items listed above, we pause nonetheless to take in the street paved with smooth river stones from the Charles River, made even smoother by years of foot and horse traffic. It's not hard to imagine Edward Pickering as a youth meandering down the length of Acorn Street, stopping to admire the haphazard pattern made by the polished river stones.

After Acorn Street, we come to the intersection of Cedar and Chestnut. The unassuming brick building on the northwest corner of the intersection contains one of the oldest music libraries of European chamber and symphonic music in the nation. It's also where members of the Harvard Musical Association have held their weekly rehearsals since the group was founded in 1837. But don't let "Harvard" in the name throw you: founding members called the group the Harvard Musical Association because many of its members were Harvard trained, but not in the music department (it didn't exist at the time).

After we linger a moment or two, we turn right and head east along historic Chestnut Street. It's historic because, along with Mount Vernon Street, it was one of the earliest streets that the Mount Vernon Proprietors developed; it's also charming with a direct view of the New

State House. Like other Beacon Hill streets, it's punctuated with gas-lit streetlamps, cast-iron balcony railings, redbrick facades, and brass doorknockers and foot-scrapers. More than that, it reflects the changing tastes and economic fortunes of the city. Between 1800 and 1807, the first houses built on Chestnut Street were freestanding mansion houses, typically four-story dwellings with side yard and carriage house and stable around back. This, after all, was the original intent of the Mount Vernon Proprietors: to build country estates for the wealthy overlooking the Charles River. That's why you'll find among the first mansions built homes for Harrison Gray Otis, Jonathan Mason, and Hepzibah Swan, all members of the Mount Vernon Proprietors. But construction halted after Thomas Jefferson signed the Embargo Act of 1807, which led to the War of 1812. As the economy began to recover after the war, home construction picked up. This wave of construction saw three significant changes: a preference for the three-story townhouse with gabled dormer rather than the four-story freestanding mansion, a shift away from Federal-style architecture to Greek Revival, and, perhaps most importantly, the growing importance of the housewright, a master carpenter who could both design and build houses. Several of these master builders put their stamp on Beacon Hill properties during this second wave of home construction, but none more than Cornelius Coolidge. In fact, by 1850, you could easily find residences designed and built by Coolidge & Company on Beacon, Mount Vernon, Chestnut, and Acorn Streets, as well as in Louisburg Square.[11]

The next intersection we encounter is at Chestnut and Spruce. Before taking a right onto Spruce Street, however, let's pause a moment and look across the street at Number 29A Chestnut Street. The townhouse occupying the lot is noteworthy for several reasons. It's the site of the first Bulfinch house erected in Beacon Hill by the Mount Vernon Proprietors (although the original structure burned down within a year of its 1799 construction). Sixty-five years later, its occupant was the great tragic actor Edwin Booth, who was staying at 29A Chestnut as the guest of its owner when he received news that his brother, John Wilkes

Booth, had assassinated President Lincoln. Shaken and unable to take the stage at Boston Theatre that night, where he was starring in *The Iron Chest* (ironically, a play about a murderer haunted by his misdeeds), Booth booked a private compartment on a midnight train bound for New York: it would be almost a year before the great tragedian would take the stage again.

We now head down Spruce Street toward the Common. Our first stop is a narrow alley on our right. Its original name, Kitchen Street, tells us that it was a service entrance for the mansions fronting Beacon and Chestnut Streets (as were most of the small lanes and alleys running east and west). Although Spruce Street was not a service alley per se, it's a fairly nondescript street, displaying the solid brick sidewalls of large multistory buildings facing Chestnut and Beacon Streets. It is at Beacon Street that our view widens, with Boston Common unfolding before us. Before turning left and heading east along Beacon Street to admire the architectural gems lining the north side of the street, let's think about what lies on the south side of the street—Boston Common.

I'm looking at a lithograph created by James Kidder in 1829.[12] It's a view of the New State House from the point of view of the Common. The viewer standing on the far side of the Common pauses to look up at the state capitol sitting majestically atop Beacon Hill. The large, imposing structure dominates the few buildings that surround it, most notably Hancock Manor, which is dwarfed by its almost temple-like next-door neighbor. Passersby stroll along tree-lined paths beneath an expansive cloud-filled sky aglow in the late afternoon sun. There's something else worth noting: in the lower left-hand corner of the lithograph, several cows gather beneath a large shade tree.

The idea of a municipal green space was a European invention, brought to the New World by early English settlers. Although the land set aside was not necessarily owned in common (the title was often held by a large land-holder or municipality), local residents—"commoners"—had access to the land for reasons of labor and leisure. Boston Common was no different. From the outset, it served a variety of interests, both public and private: couples took afternoon strolls along tree-lined malls, children played in eroded battlements left over from British occupation, workers dug clay and gravel and removed stones for building purposes, local militias drilled in lockstep, women gathered to wash clothes, men gathered to converse (and, occasionally, to duel), officials staged well-attended public executions, and cows grazed leisurely beneath the spreading arms of towering elm trees, which confirms what Michael Rawson, author of *Eden on the Charles*, notes: "In Boston Common's first two hundred years, there were few kinds of activities that it had not accommodated at one time or another."[13]

By the early nineteenth century, however, most of these activities either had been banned or had been moved to another part of Boston (this included the municipal buildings at the eastern end of the Common, which included an almshouse, prison, workhouse, and granary). The most contentious debate, however, involved banning cows from the Common. The grazing of cows on commonly held land had been a cherished right of Boston's earliest denizens. The city's selectmen understood this, but they also understood their responsibility to protect the city's pastureland from overgrazing in order to preserve the Common for its many other purposes. As such, over the years, they restricted residents to the grazing of one milk cow per family from early spring to fall (while levying a tax for the privilege of doing so), while banning horses, sheep, goats, and "dry" or non-dairy-producing cattle.

By 1830, however, the area surrounding the Common had become a prime location for high-end residential development and new homeowners began calling for an outright ban on all dairy cows grazing on the Common. Although many Bostonians protested the loss of this long-held right, in the end, they could not fight the power and influence of Boston's upper-class families who were steadily moving into homes that fronted the Common along Beacon, Park, and Tremont Streets. Ironically, the man behind the expulsion of the cows from the Common was none other than lawyer-turned-land-speculator-turned-politician Harrison Gray Otis, who at the time was serving as Boston's third mayor and living in a lavish mansion overlooking the Common.

Although it's easy to imagine Edward Pickering walking through the winding paths of the Common on a sunny afternoon, it's just as easy to imagine him trundling along Beacon Street rattling a stick along the ornamental cast-iron fence that encloses the Common. To get an on-the-ground view of such a scene, let's study an engraving by John Andrew, a Boston engraver who supplied images to the publisher Ticknor and Fields before opening his own firm, John Andrew & Son. The engraving, created between 1846 and 1849, is a panoramic view of Beacon Street looking west titled "A View of Beacon Street, Boston."[14] On the left-hand side of the engraving, sinewy trees tower above the tall, handsome fence that encloses the Common. On the right-hand side of the image, a line of single-family dwellings—some freestanding, others clustered together—face the Common. Given the proper attire of the people on the street, the scene appears to take place on a Sunday afternoon: stylishly-dressed couples cross the cobblestone street on their way to the Common; four-wheeled horse carts ferry cheery-eyed occupants from one place to another; boys in Sunday suits accompany stone-faced adults down the wide brick sidewalk; and nearby, men with ivory-tipped walking sticks chat amicably with each other. It's a bucolic, pastoral scene: a scene much different from the bustling wharf district near the city center.

As we turn left and head east on Beacon Street, what Oliver Wendell Holmes called "the sunny street that holds the sifted few,"[15] we stop at several residences of note. First up is Harrison Gray Otis's grand Federal-style mansion at 45 Beacon Street. Built in 1806, the four-story stone-and-brick structure was the third mansion that Bulfinch built for Otis in Beacon Hill within the span of a dozen years. One of the "sifted few," Otis moved into the thirty-seven-room house, with its eleven bedrooms, ten bathrooms, and twelve fireplaces, with his wife, six children, and half a dozen servants.

The structure is impressive: a freestanding mansion set back from the sidewalk, with formal gardens, a cobblestone driveway, and a carriage house with horse stables nestled at the back of the property. Here's how noted historian and Boston walking tour guide A. McVoy McIntyre describes the mansion:

> Originally the structure stood free, with its east side gracefully bowed to form the oval room so admired of the period. This gave onto a spacious garden which, graded into the slope of the hill, was a level above the street entrance. The prominent feature of the front façade is the piano nobile, expressed here with flawless grace. The deep windows, defined in a classic wood framing with bracketed lintel, once commanded a splendid view over the green pasture of the Common and across the waters of Back Bay to the Blue Hills.[16]

At ten years old, Edward Pickering probably wouldn't stop to appreciate the finer details of the Otis mansion, which also included a rectangular portico flanked by delicate pairs of Ionic fluted columns that framed the first-floor entrance. I'm sure this would be of little interest to Edward. A ten-year-old boy would be more interested in peeking through the first-floor window to marvel at the ten-gallon blue-and-white Lowestoft punch bowl sitting at the bottom of the grand staircase, strategically placed there in case one of Otis's visitors needed to quench his thirst before ascending the staircase to the main receiving room.

After leaving Number 45, we pause in front of 42 Beacon Street, the home of Otis's neighbor, Col. David Sears. The difference between Otis's redbrick, Federal-style mansion and Sears' granite-faced, Greek-Revival home is striking, with the latter more appropriate for an English baron than a successful Boston merchant. If Edward Pickering were passing the structure in 1856, he would have seen a two-story, freestanding mansion with a single domed bay faced with Rockport granite. Designed by Alexander Parris, Sears built the mansion in 1819 to reflect many of the elements of the Mount Vernon Street estate of his father-in-law, Jonathan Mason. Like Otis's mansion, Sears' home was built as a country estate with formal gardens, carriageway, and coach yard and stables in back.

Resuming our tour, we continue on until we reach the northwest corner of Beacon and Walnut Streets. The address should be 38 Beacon Street, but it's not, since the building's entrance is on Walnut Street (although it's unclear when the entrance was changed). Built in 1804 for John Phillips, Boston's first mayor and the father of noted abolitionist Wendell Phillips, the structure was the first brick house built on Beacon Street. Although originally built in the Bulfinch Federal style, over the years, the Phillips' house has seen several Victorian additions, most notably a heavy-looking mansard roof. When Edward Pickering first walked by the house, the Phillips' family would no longer live there: it was sold in 1825 to Lt. Gov. Thomas Winthrop, a descendant of John Winthrop, the Massachusetts Bay Colony's first colonial governor.

As we ponder this, we come to the last property on Beacon Street worth noting before we turn north on Joy Street: Number 34½ Beacon Street. As noted earlier, Joy Street (formerly Belknap Street) was named after Dr. John Joy, a noted Boston apothecary who bought a large tract of land in 1791 west of Belknap Street that belonged to the estate of John Singleton Copley. Dr. Joy built a modest two-story house on the northwest corner of Beacon and Belknap, selling off land to the north in order to regain his initial investment. Dr. Joy and his wife lived at the corner of Belknap and Beacon until his death in 1806, at which time the property was sold and subdivided. Uriah Cotting, one of the great land developers of the early nineteenth century, bought the corner lot in 1806 with the intention of erecting one of the largest mansions that Boston had ever seen.

Unfortunately, Cotting, like many of his wealthy friends, suffered the financial repercussions of Jefferson's Embargo Act of 1807 and was forced to sell. Financier Israel Thorndike, one of the richest men in America at the time, bought the property in 1810 and promptly erected a mansion that quickly became the social and political nexus for Boston's elite. After Thorndike died in 1832, Robert Gould Shaw, a wealthy merchant and grandfather of Col. Robert Gould Shaw who led the 54th Regiment of free blacks during the Civil War, bought the property and lived there until his death in 1853, whereupon Frederic Tudor, known as the "Ice King," bought the property and erected a mansion of his own.[17] In 1856, Edward Pickering would have passed—and admired—the Tudor mansion, a three-and-a-half story building with granite base and brick walls. Other distinctions include two bays that faced Beacon Street enhanced by first-floor iron railings, prominent stone lintels and sills on every window, elegant brick chimneys, and a Greek-Revival entrance off Joy Street.

Edward would see other Greek-Revival structures as he continued north along Joy Street, once just an alley for carting hay to the Common's pastureland, until he arrived at the intersection of Joy and Mount Vernon Streets. On the southeast corner of the intersection, Edward would see a double building; not a twin townhouse, but two houses of different sizes connected to each other. Alexander Parris designed and built the four-story corner structure in 1824 for George Williams Lyman, a successful businessman and scion of a prominent Boston family. After purchasing the building adjacent to the corner property, Lyman connected the buildings with a full enclosure and turned the complex into a family compound that stayed in the family for generations. The compound also sported a large garden area—once the site of John Hancock's garden—that faced Joy Street and Mount Vernon Place, a small alley running behind the Lyman property.

We now come to the last leg of our Beacon Hill tour—Mount Vernon Street. As we begin to walk the length of the street, we are struck by what greets us: a solid wall of townhouses, punctuated occasionally by a few remaining freestanding mansions. We'll pass by the first half-dozen structures and stop in front of Number 55 Mount Vernon Street—the Nichols House (named after Rose Standish Nichols, world traveler, author, and founder of the Women's International League). But Nichols was a late-comer to the scene: 55 Mount Vernon Street was originally part of the multi-lot family compound of Jonathan Mason, who built a mansion for himself (lots 59-67) and townhouses for each of his three daughters (lots 51-57).

What strikes us as we stand on the sidewalk in front of the Nichols House is the thirty-foot setback from the sidewalk of each house as we look west from Number 55 all the way to Number 89. This was the result of a gentleman's agreement between two Mount Vernon Proprietors—Jonathan Mason and Harrison Gray Otis. With Mason's compound between 51- 67 Mount Vernon Street and Otis's mansion at 85 Mount Vernon Street, the two gentlemen (if real estate tycoons can be called "gentlemen") agreed to set each building back thirty feet in order to keep a straight line of view down the street. The year was 1801 and they owned the street—literally. Well, almost all of it: Charles Bulfinch owned several lots between the two complexes, but due to another round of financial setbacks, he was forced to sell them.

Continuing on, we cross Mount Vernon Street to take a closer look at several structures on the south side of the street that seem to be somewhat out of place. Rather than elegant townhouses, the structures in question look more like garden sheds. And you'd be wise to think that since they were once the stables at the back of properties fronting Chestnut Street built by heiress Hepzibah Swan for her daughters. By deed, the stables at the back of the lots were limited in height to thirteen feet. Aside from the low building height, we quickly notice four doorways, each one slightly recessed into the brick wall: three single-wide doorways capped by a horizontal bar of glass transom panels and one double-wide doorway capped by a fanlight transom with handsome radial lines. If you stood here long enough in the early 1800s, you'd see what the fourth doorway was for: it was a stable door that allowed a horse-and-carriage access to the interior coach yard and stables.

Let's re-cross the street and stop in front of Number 85 Mount Vernon Street—the second mansion that Harrison Gray Otis built in Beacon Hill. Unfortunately, we have to strain to see the front façade in its entirety, as the yard is raised at least four feet from street level and the stone retaining wall and accompanying iron fence is backed by a line of small trees and bushes. But if we shift our location to the mouth of the cobblestone driveway that runs along the east side of the building (to a side door and then to the coach house around back), we can see the building in all of its glorious detail. Built in 1802, the Bulfinch-designed home is a large square Federal-style structure, three-stories in height with brick walls laid in Flemish bond. Although the structure echoes elements that Bulfinch used in Otis's first mansion built in 1796 on Cambridge Street, the Mount Vernon Street structure has many distinguishing elements of its own: recessed first-floor windows ornamented with Chinese fretwork balconies, windows framed by black shutters that diminish in size with each succeeding floor, a broken entablature over the third floor supported by Corinthian pilasters that span the top two floors, and a roof balustrade and octagonal cupola that lends an elegant touch to the imposing structure. And, believe it or not, Otis and his wife, Sally, lived in the home for only four years. Otis already had his eye on another piece of property for his growing family—45 Beacon Street.

We finally arrive at our last piece of property to examine before returning to Edward Pickering's childhood home at 74 Mount Vernon Street. For this piece of property, we'll merely turn around and look across the street. You really can't miss it, given the abrupt change of style. I'm sure Edward and his neighbors wondered about it as well. The address is 70-72 Mount Vernon Street. The massive Italianate structure is comprised of two mirror-image townhouses on property just east of the Pickering's modest three-story brick townhouse. It was built in 1847, the year after Edward Pickering was born, on the granite foundation of an older structure: the coach house and stables of a mansion that faced Chestnut Street. For that property, Richard Derby of Salem, Massachusetts, hired the ever-popular Charles Bulfinch to design a freestanding mansion and accompanying coach house and stables. Why Bulfinch? Because the Derby family mansion in Salem, though built by the talented housewright Samuel McIntyre, was based on sketches provided by Bulfinch. Although Derby's Chestnut Street mansion—Number 27— was one of the most elegant in Beacon Hill at the time, it was torn down in the mid-1840s to make room for two other buildings—one fronting Chestnut Street; the other, Mount Vernon Street.

It is the building that fronts Mount Vernon Street, the twin Italianate townhouses, that concern us here. The Thayer brothers—John and Nathaniel—who amassed a fortune in banking and railroad investments, commissioned Richard Upjohn, an English architect working in New York City, to design the building. Upjohn had just completed Trinity Church and was involved with enlarging the Hudson River estate of Stephen Van Rensselaer, whose daughter had married into the Thayer family. Originally, the Italian Palazzo-style building, one of the last paired houses built in Beacon Hill's South Slope neighborhood, was erected as a three-story stone structure, six bays wide, with a steep gambrel roof with dormers (that was what Edward Pickering would have seen when he passed it as a youth). The façade drew heavily from elements of the Italian Renaissance: two slightly projecting outer bays acting like pavilions contained an arched doorway framed by stone pilasters and protected by decorative stone balconies. Completing the integrity of the façade are vertical quoins extending the height of the building, a guilloche-patterned stringcourse running between the first and second floors, and pediment-looking stone lintels capping each of the third-floor windows. (Before Pickering reached his twenty-fifth birthday, the building would experience several notable changes: in 1870, the roof was removed in order to add two stories to the already massive structure; additionally, a bank of large windows was inserted on the second floor to allow more sunlight into the library.)

As we digest the unique elements of this building, let's turn and head west along Mount Vernon Street. But not far; we'll stop at the first townhouse we come across, which is where we began our tour—in front of Edward Charles Pickering's childhood home at 74 Mount Vernon Street. Located next to the Thayer building, Edward's childhood home is both unassuming in appearance and, given the building next door, diminutive in size. But it's not the structure that distinguishes the Pickering family home: it is—and always has been—the address. Number 74 Mount Vernon Street functions as a kind of family coat of arms, signaling to the world the family's rising fortunes. Although never considered legitimate members of the Boston Brahmin upper class, Edward and Charlotte Pickering could hold their heads high as they walked the streets of Beacon Hill's South Slope. They had earned the right to do so, through their family lineage, Edward's hard work, and the education Edward and Charlotte sought for their children.

It is to that education, especially their elder son's education, that we now turn. Since no definitive records exist to help us identify the first school Edward Charles Pickering attended, we'll have to make our best educated guess. We do know from records kept by the Boston Latin School Association that Edward Pickering attended Boston Latin from the fall of 1856 to the spring of 1862, graduating at the age of sixteen. What we don't know is where he went to school prior to Boston Latin. Several biographical sketches suggest a "private school." But public or private, we really don't know precisely, so we'll have to do some detective work in order to narrow it down.

CHAPTER 2 | *Reading, Writing, Refmetic*

> *Among the most important events in a boy's life, the admission to his first "man's" school stands forth with prominence. I remember the feeling of pride and awe which accompanied me upon the morning of my debut. It was a good school, and was kept by Mr. Franklin Phelps, who died but a few years ago, very aged, and, I believe, totally blind.*[1] —James D'Wolf Lovett

It's fortunate that Edward Pickering grew up in Boston during the middle of the nineteenth century, since Massachusetts was one of the leading states that grappled seriously with the education of its youth. The origin of this concern comes from one of the first European groups to immigrate to the New World. They were, as the history books tell us, "dissenters"—Puritans—who fled England in order to escape persecution from the king and leaders of the Anglican Church. Deciding that the church was too corrupt to reform, they arrived in the New World ready for religious, moral, and social reform. The basis of Puritan reform lay in their belief that the Bible was God's true law. In their attempt to purify the church, they stripped away many of the formalities of traditional Christianity. In doing so, Puritan leaders, who held strong personal and religious beliefs, created a strict, fundamentalist religious environment in which—ironically—dissent was not tolerated: freethinkers were whipped, apostates exiled, and heretics burned at the stake.

They did this not only to maintain religious purity, but also to survive the hardships they encountered in their newly adopted land. A strong sense of community was one of the best defenses against threat, both from within the community and without. Any deviation from the tenets set out by church elders was met with fierce disapproval and discipline. And since church elders were also the community's political leaders, any infraction against the church was an infraction against the entire group. The secular and the religious were woven together in a tight-knit fabric.

Despite their strict social and religious views, Puritans brought with them a penchant for education, ardently supporting the means of learning—books, libraries, printing presses, and schools. They did so not out of a spirit of progressive liberalism, but out of a belief that the uneducated—especially in religious matters—could easily fall prey to the devil. For Puritans, the devil was behind every evil deed, and since men and women—and children especially—were inherently bad, religious instruction was essential in order to learn how to behave. This worldview placed education at the heart of the Puritan experience, with the home the first line of defense.

Puritans lived a strict, disciplined life centered on religious study and manual work. There was little time for leisure. What little free time children had was usually filled with instruction: learning the basics of reading and writing, which meant memorizing the alphabet, reading the Bible, saying simple prayers, and learning how to write one's name. There was also "ciphering," or simple arithmetic. It was home schooling in its purest form: the family gathered at night, with instruction meted out in the strictest manner by the head of the household—in most cases, a stern father figure.

Over time, as the population of the colonies swelled in number, other informal means of education arose for young children. One form was the Dame school, which was set up in the home of a woman (often a widow or "Dame"), who had both the time and inclination to care for and tutor young children. Children—boys and girls up to the age of eight—were taught the letters of the alphabet, spelling, grammar, and simple arithmetic, their parents paying a modest fee for the service. Attendance was erratic at best, often depending upon the seasonal work cycle. In either case, children learned at home until the age of eight, with one of three options awaiting them: they—"they" meaning boys of European descent—either entered the town school (run by a local clergyman or an itinerant teacher), remained at home to learn their father's trade, or apprenticed with a local craftsman. Girls, on the other hand, had none of these opportunities: they continued their education at home as their future work was in the home, raising a family and running a household.

The close-knit, informal nature of instruction in the early days of the colonial period served the needs of the community well, as long as the people in those communities shared the same set of beliefs. But less than a half century after their arrival in the New World, church elders felt that the Puritan way of life was being threatened. Not by internal dissenters, but by new waves of immigrants, non-believers, heretics with different religious beliefs. In order to maintain the Puritan ethos, the masses—especially non-Puritans—had to be educated in the Puritan way in order to understand the written codes—both religious and secular—that people in the colonies were expected to follow.

The result was passage of the Massachusetts School Law of 1642, the first compulsory education law passed in the colonies, which placed the burden of educating children squarely on the shoulders of parents. It required that parents (and masters of children apprenticed to them) saw to it that children knew the basic principles of religion and the capital laws of the Commonwealth. The idea behind the Massachusetts School Law of 1642 was simple: if citizens could read and write, then they would be able to understand—and, therefore, abide by—the governing laws of the land, both religious and secular.

Fearing that parents and masters weren't taking the law of 1642 seriously, the Massachusetts General Court passed the General School Law of 1647. The law, also known as the Old Deluder Satan Act, required towns of at least fifty households to hire a schoolmaster to teach children to read and write, to do simple arithmetic, and to learn basic principles of religion. The Old Deluder Satan Act was quite clear in its message:

> Every township in this jurisdiction after the Lord hath increased them to fifty households shall forthwith appoint one within their town to teach all such children as shall

> resort to him to write and read, whose wages shall be paid either by the parents or masters of such children, or by the inhabitants in general, by way of supply.[2]

Towns of at least one hundred households were given an additional responsibility: they were to establish a Latin grammar school in order to prepare students—that is, boys from the upper classes—for entry into college where they would prepare for the ministry or learn a profession.

One effect the passage of these laws had was to increase the need for teachers, who, in the earliest days of the colonial experience, were most often ministers of local congregations. But with the passage of compulsory school laws, a growing need arose for more teachers who could teach students in small one-room schoolhouses that were beginning to appear in every town and hamlet in Massachusetts. No longer was the education of young children a matter of the home; after 1647 the responsibility came more and more under the purview of society. Teachers were hired—and schoolhouses built—for the sole purpose of teaching young people in every Massachusetts community.

The next law that impacted the education of children in the Commonwealth came in 1701 when the Massachusetts General Court increased the fine for failure to maintain a schoolmaster from ten to twenty pounds sterling. The fine was substantial enough that many towns that had earlier paid the ten-pound fine rather than hire a schoolmaster complied with the law, seeing it as the less expensive of the two alternatives. However, the law had an unforeseen consequence on town schools. Many townships that covered a large area had a difficult time raising public funds to support a town school, as residents living some distance from the proposed school had little interest in supporting a school to which their own children could barely attend. As a result, a new form of town school arose, the "moving school," which was kept for part of the year in each of several sections of the township, as this 1702 entry in the town record for Malden, Massachusetts, reflects:

> John Sprague is chosen schoolmaster for ye year ensuing to teach children to read and write and to refmetic according to his best skill, and he is to have ten pounds paid him by ye town for his pains. The school is to be free for all ye inhabitants of this town: and to be kept in four several places, one quarter of a year in a place.[3]

The problem with the moving school, however, was that while it gave *every* child an opportunity to attend school at some point during the year, it lessened the overall amount of instruction time for *each* child. At the time, it seemed a necessary arrangement in order to entice citizens of a large township to vote for a public school tax. As the population of the Commonwealth of Massachusetts increased, maintaining one schoolmaster per township—as required by the law of 1642—became impractical. As a result, many populous townships, especially those spread across large tracts of land, created district schools within the borders of their township. Instead of spending township money on the maintenance of one good school to which some children traveled a great distance (or a moving school of lesser duration), each district within a township was granted part of the tax dollars collected by the entire township, applying it toward the maintenance of a smaller district school. Although an economically sound move, the practice only escalated the decentralization of school authority, often leaving local district schools in the hands of unprincipled, self-serving individuals.

The Massachusetts General Court institutionalized the practice of dividing up townships into smaller district schools in 1789 when it passed a law legalizing the custom. In no time, smaller districts were made corporations in their own right with the power to enforce contracts, levy taxes, and, by 1827, choose a committeeman who had charge of school property and the hiring of teachers. The law also changed the amount of continuous time a school had to be in session, reducing it to six months; it also reduced by half the number of Latin grammar schools previously required (instead of one school per one hundred families, the new law required one school per two hundred families).

Perhaps even more devastating to the quality of local public education, the law also changed the qualifications of schoolmasters. Instead of a master of "good morals, well instructed in the Latin and Greek and English languages," a teacher had to be "well qualified to instruct youth in Orthography, Reading, Writing, Arithmetic, English Grammar, and Geography, and in good behavior...."[4] To those families who had hopes of sending their sons across the Charles River to Harvard, or abroad to Cambridge University, the change was disappointing, as it made it harder for boys to meet college entry requirements, which were eloquently stated in 1642 by Henry Dunster, Harvard's first official president:

> Whoever shall be able to read Cicero or any other such like classical author at sight and make and speak true Latin in verse and prose, *suo ut aiunt Marte*, and decline perfectly the paradigms of nouns and verbs in the Greek tongue: Let him then and not before be capable of admission into the college.[5]

The result, a trend that had been steadily growing from the earliest days of the colonies, was the spread of private academies designed as college preparatory schools for the sons of wealthy New England families. Such academies appealed to those of means who looked down on the common district school. Private academies, often endowed by public-spirited individuals and just as often supported by land grants from the state, flourished after the Revolutionary War period, so much so that by 1825, there were nine hundred and fifty private schools in the Commonwealth of Massachusetts alone.

It was evident that by the early nineteenth century, a two-tiered system of education was evolving: one that adequately addressed the needs of the wealthy, and one that barely addressed the needs of the poor. The growing rift between public and private education began to alarm educational reformers who believed in the principle of the "common school," free and open to all of the Commonwealth's young citizens. Massachusetts state legislator James Carter was one of those reform-minded individuals alarmed by what he saw. In 1826, he began a series of attacks on the perceived ills of Massachusetts' educational system, railing primarily against the decentralization of school authority, which he believed was at the heart of the Commonwealth's under-performing public schools.

One of the first laws that Carter helped shepherd through the state legislature required every Massachusetts town to choose a school committee to oversee the town's schools, the selection of textbooks, and to examine and certify its teachers throughout the town (though it left it up to each district to hire its own teachers). It was a significant law, passed in 1826, in that it reversed the direction of school decentralization by placing public school supervision within a township under a single authority.

Eight years later, again with Carter's leadership, the Massachusetts General Court created the nation's first state school fund that required towns—in order to receive a portion of the fund—to levy a tax for each school-age child and to make annual school reports on the operation of its schools and the progress of its students. Carter's efforts culminated in 1837 with the passage of a law that established the nation's first State Board of Education. The board consisted of eight members appointed by the governor and members of a legislative council. Although the board had no executive power, its members were charged with gathering information concerning the state's schools along with recommending appropriate changes to school organization, supervision, and instruction.

To lead the Board of Education, legislators turned to one of their own—Horace Mann, who at the time was serving as president of the Senate. Mann, a well-trained and successful lawyer, gladly gave up his presidency and accepted the position of the state's first Secretary of the State Board of Education, a position he held for the next twelve years. A person of high moral principles, and one thoroughly devoted to public service, Mann worked long and hard to rekindle the Common School Movement. He firmly believed in the concept of common schools, schools open to all students of all walks of life, boys and girls, which was—after all—the original intent of the Massachusetts school laws passed over a hundred years earlier. According to Mann, the common school was the "great equalizer" that put all men and women on the same educational footing.[6]

Knowing that the Massachusetts General Court had not endowed him with executive authority, Mann used his powers of persuasion—through personal addresses and detailed annual reports—to convince residents of the Commonwealth of Massachusetts that a strong, centralized public school system was in the best interest of everyone, rich or poor. His commitment to universal public education solidified after a trip to Europe. While on this trip, Mann fell under the influence of the Prussian model of common schools, which held that every child was entitled to the same teaching practices and instructional content. Upon his return, Mann prompted fellow educators, state legislators, and concerned citizens to throw all of their weight behind the Common School Movement.

Along with his relentless support of a public school system funded by public tax dollars, Mann lobbied to keep religious education separate from public education. He introduced age grading, in which students were assigned by age to different grades, a practice that almost destroyed the one-room schoolhouse overnight. He argued for a longer school year. He sought improvements to the school curriculum. He convinced the state legislature to increase expenditures on education, which it did significantly. At the root of all of his efforts, Mann believed that the key to a successful public school system lay in one thing—the quality of its teachers. As such, Mann worked tirelessly to create a network of well-trained teachers, through the development of teacher training colleges called Normal Schools, that could deliver a common education to all children residing within the Commonwealth of Massachusetts. According to Ellwood Cubberley, no one did more than Horace Mann to establish in the minds of the American people the idea that "education should be universal, non-sectarian, free, and that its aims should be social efficiency, civic virtue, and character, rather than mere learning or the advancement of sectarian ends."[7]

But Mann's efforts did not come without resistance. Although many groups opposed Mann's efforts based on philosophical differences, just as many opposed his efforts out of pure self-interest. Taxpayers worried that public schools would take money out of the pockets of the working-class in order to pay for the education of the rich. Church groups worried that public schools would fail to teach religion sufficiently. Private schools worried that public schools would siphon off the best-trained teachers. Although harsh in their criticism, they succeeded in slowing but not stopping Mann's call for free, public, tax-supported education.

By the time Mann stepped down as Secretary of the Massachusetts State Board of Education in 1848 to fill the congressional seat vacated by the death of John Quincy Adams, an organized system of public education was beginning to take shape throughout the Commonwealth of Massachusetts, a system bolstered by passage of the nation's first compulsory school attendance law in 1852 that required school-age children in Massachusetts ages eight to fourteen to attend school three months out of the year, with at least six of the twelve weeks attended consecutively. In short, free, tax-supported, centralized public education was here to stay.

IN THE SAME YEAR that the Massachusetts General Court passed the nation's first compulsory school attendance law, Edward Charles Pickering turned six years old, two years younger than the compulsory school attendance age, but old enough for his parents to begin planning for their eldest son's formal education. Like his recent forebears, who lived in or near Beacon Hill's South Slope neighborhood, Edward Pickering would be well educated. There was no doubt about that. Edward would attend Boston Latin and then matriculate at Harvard College, where presumably, he would study law. It is the first step, however, of which we are uncertain: Just where did young Edward go to school prior to Boston Latin? It may not seem important to know the answer to this question, if it can be answered at all, but by looking into it, we just might learn more about Boston's evolving educational system.

The first organized public school experience for children in Boston in the mid-nineteenth century was typically the local primary school, which took in children ages four to eight years. Under the law of 1789, illiterate children were barred from grammar school attendance, which began generally around age eight. If parents couldn't teach their children the basics of literacy at home, then they had to send them to a private school for such instruction; that is, if they could afford to do so. When the first public primary schools began appearing in Boston after 1818, under the direction of the newly created Primary School Committee, the burden of teaching their children was lifted for many families who were either unable to teach their children or couldn't afford to send them to private school.

Like private nursery schools—and, later, public kindergartens—primary schools hired a female teacher to teach the basics of reading, writing, spelling, the use of the slate, simple arithmetic, and sewing. At age eight, each student—if deemed qualified—received a certificate of transfer to the local grammar school. The transfers took place semi-annually on the first Monday in March and in July when school broke for summer vacation. Children were given monthly, quarterly, and yearly exams (the last of which was administered by members of the Executive School Committee during the first two weeks of May).

Since illiterate children were barred by law from entering grammar school, children who reached age eight and had not passed the literacy and numeracy tests had to continue their education at a special school, called an "intermediate" school. Once students passed their exams, they entered grammar school. However, the grammar school of nineteenth-century Boston is not to be confused with the Latin grammar school of seventeenth-century Boston, which was modeled after the English classical school that emphasized the mastery of Latin, Greek, English literature, religious history, and math. By the mid-nineteenth century, Boston grammar schools had evolved (or devolved, depending upon your point of view) into a system of public elementary schools that barely resembled the original English classical liberal arts model.

In fact, by the end of the nineteenth century, the term "grammar school" had all but disappeared in preference to the more general term "elementary school." In other words, although the growing number of grammar schools in 1850 Boston might have served well the general school-age population, they did not serve the specific needs of elite, upper-class families intent on sending their sons to college. And Edward Pickering's parents were intent on just that, which meant that Edward and Charlotte Pickering would not have sent their eldest son to a local grammar or elementary school, since its curriculum, admirable for the time, would not have adequately prepared Edward for entry into Boston Latin. Realistically, there was only one choice for Edward and Charlotte Pickering when it came to choosing a school for their eldest son. Edward had to attend a private school taught by a certified Latin grammar master. There was really no other way for Edward to obtain the requisite education he would need to pass the entrance exams to Boston Latin, which, in turn, would secure him entry into Harvard College.

In all likelihood, then, Edward stayed home under the supervision of his mother, a live-in relative, or governess, until he was six or seven years old, at which time his parents would have enrolled him in his first "man's school,"[8] a local private school capable of preparing him—and other boys of European descent—for the rigorous entrance exams to Boston Latin. Since this was the age of the horse and buggy, more than likely, Edward Pickering attended a school within walking distance of his home on Mount Vernon Street—but which one?

In his reminiscence about living in Boston during the middle of the nineteenth century, James D'Wolf Lovett gives us a clue as to which school Edward might have attended: "Private schools in those days were not too plentiful, and the best were Phelps's, Sullivan's, under Park Street Church, Prescott Baker's, in Chapman Place, and Mr. Dixwell's Private Latin School, in Boylston Place."[9] Let's take a look at each of these schools, plus Chauncy Hall School, and try to determine which school young and impressionable Edward Charles Pickering might have attended.

Dixwell School. Born to Boston physician John Dixwell and Esther Sargent Dixwell, Epes Sargent Dixwell took the usual course of studies that most sons of prominent families did: he attended Boston Latin, then Harvard College—and all before his sixteenth birthday. He was a natural scholar with interests in literature and music. After graduating from Harvard in 1827, he taught at Boston Latin as a sub-master and part-time at the English High School. In 1830, Dixwell abandoned his education career to study law at Harvard, apprenticing with Charles Loring, one of the most eminent lawyers in Boston. Dixwell passed the Massachusetts Bar in 1833 but gave up his practice three years later to accept the position of headmaster at Boston Latin.

A successful, well-admired, but feared schoolmaster, Dixwell stepped down fifteen years later because of a residence requirement passed in 1851 by the Boston City Council. Nine years earlier, Dixwell and his wife had moved to Cambridge, preferring the town's quiet, tree-lined streets to Boston's busy thoroughfares, as well as Cambridge's more intellectually stimulating environment. However, after the Boston City Council passed an employee residence requirement, Dixwell was given a choice: resign or move back to Boston. Dixwell decided to resign, and then turned around and opened his own private Latin grammar school in the heart of Boston, naming it The Private Latin School of Boston (although locally, it was known simply as Dixwell School). In *Old Boston Boys and the Games They Played*, Lovett portrays the school as follows:

> Mr. Dixwell's school, founded in 1851, was located in Boylston Place, in a house which he built specially for it, and was recognized as the finest private school in Boston. Boys were here fitted for college, and any graduate who bore Mr. Dixwell's hallmark was sufficiently guaranteed without further question.[10]

Ernest Samuels, writing about Henry Adams' experience at Dixwell School, noted that Dixwell prepared young scholars "with single-minded diligence for the Harvard entrance examinations, studying Latin, Greek, mathematics, history, geography, English composition, and, of course, declamation."[11] One of the more detailed descriptions of Master Dixwell comes from Massachusetts State Senator Henry Cabot Lodge, who wrote in his memoirs:

> Mr. Dixwell was a direct descendant of John Dixwell, the regicide, who sensibly took refuge in Connecticut when the estimable Charles II came to the throne. I have thought since, perhaps fancifully, that a certain stiffness and rigidity which were observable in my master, who was a good deal of a martinet and given to severe sarcasm at the expense of stupid or disorderly boys, may have ben inherited from his conspicuously Puritan ancestor, who had passed sentence of death upon a king. But what I never doubted was that Mr. Dixwell was a thorough gentleman, albeit a rigorous one, and that he was also a scholar and an accomplished man.[12]

He must have been accomplished, and his instructional methods must have worked, because Dixwell obtained consistent results: between 1851 and 1872, the year the esteemed schoolmaster retired, almost five hundred scholars matriculated at Harvard. Regarding his methods of instruction, Senator Lodge added:

> We spent a great deal of time on the Latin and Greek grammars and mastered them thoroughly. We learned to read and write Latin and to read Greek with reasonable ease, going as far as Virgil, Horace, and Cicero in the one and in the other concluding with Felton's Greek Reader, which contains selections from nearly all the principal poets and prose-writers of Greece.... In addition to the classics we were drilled in algebra and plane geometry, and were given a smattering of French as well as courses in Greek and Roman history.[13]

It is precisely because of the success of Master Dixwell that we can rule his school out. If Edward Pickering had attended Dixwell School prior to attending Boston Latin, which we know he did, why would Edward's parents transfer him to Boston Latin if Master Dixwell could have prepared Edward for entrance into Harvard directly? That seems a bit illogical. Therefore, barring some fallout with Master Dixwell (of which we have no evidence), we have to assume that Edward Pickering never attended Dixwell School.

Chauncy Hall School. Like Dixwell School, Chauncy Hall School was a highly respected college preparatory school located at the southeastern tip of Boston Common. Founded in 1828 by Gideon Thayer, a noted pioneer in the field of education, and named after Dr. Charles Chauncy, minister of the First Church, which stood adjacent to the school on Chauncy Place, Chauncy Hall School trained the sons of well-to-do Bostonians for careers in business, law, and medicine. Like Dixwell School, it too was a full-service institution with a Preparatory Department (primary school) for younger boys and an Upper Department (secondary school) for older boys.

In 1868, the school moved to an old family residence on Essex Street, where it remained until the summer of 1873, when a fire destroyed the school and its library. The following September, Chauncy Hall School opened on Boylston Place, between Clarendon and Dartmouth Streets, across from Copley Square in full view of Trinity Church, where it remained for many years.

Before the school moved in 1868, however, Chauncy Hall would have been a good choice for Edward Pickering since, like Dixwell School, he could have received a first-class education which would have fully prepared him to matriculate at Harvard. However, Edward Pickering did not attend Chauncy Hall School. We know this because of a sketch written by Thomas Cushing, Chauncy Hall School's second headmaster, which includes a list of students who attended the school between the years 1828 to 1894. The list includes several Pickerings, namely Arthur Pickering and McLaurin Pickering, both of whom are listed as "original stockholders" in the 1875 directory; and John Pickering, Edward's well-known grandfather. However, there is no mention of Edward Charles Pickering.

Prescott Baker School. Prescott Baker School was located on Chapman Place, between Bromfield and School Streets, due south of King's Chapel and its burial ground on Tremont Street. Little is known about the school, or even about Prescott Baker. Both *The Boston Directory* and the *Harvard Alumni Directory* list an Amos Prescott Baker, with addresses on Derne, Pinckney, and Newbury Streets respectively.

Two other references exist that might shed light on Amos Prescott Baker and Prescott Baker School. *The New England Historical and Genealogical Register* states that one Edmund Dearborn was a teacher at Chapman Hall School, with an additional note that Amos Baker was the school's principal at the time. Is this the same Amos Prescott Baker mentioned in *The Boston Directory* and *Harvard Alumni Directory?* Furthermore, is Chapman Hall School another name for Prescott Baker School, which at the time was located on Chapman Place? Without any corroborating evidence, it's hard to tell.

But it might be a moot point anyway. In *Building Victorian Boston: The Architecture of Gridley J. F. Bryant,* Roger Reed includes an annotation in a section on Boston fires that indicates that a building belonging to Amos Baker on Chapman Place was destroyed by fire in 1852, when Edward Pickering was six years old. And what does that mean? What it means is that if there was no building, there was no school, and if no school, no students, and if no students, then no Pickering. And that leaves two schools, both within walking distance of Edward's home on Mount Vernon Street, that the young scholar might have attended: Sullivan School on Park Street and Phelps School on Chestnut Street.

Sullivan School. More is known about Park Street Church at the corner of Park and Tremont Streets than about the small school that operated in its basement in the mid-1850s. The school, run by former clergyman Thomas Russell Sullivan, took in students ages five and up, preparing them for local fitting schools, schools whose sole purpose was to prepare young scholars for entry into Harvard College. Sullivan, an accomplished schoolmaster, descended from a distinguished Boston family: he was the grandson of James Sullivan, a former governor of Massachusetts, and the grandnephew of John Sullivan, a distinguished

Revolutionary War general. Beyond this, the historical records don't say much, except to mention that scholars often transferred from Sullivan School to Dixwell School to finish their grammar school education before entering Harvard.

In today's lingo, we would call Sullivan School a "feeder school," a primary school that prepared young boys for their secondary school education. The school it prepared them for appears to be Master Dixwell's Latin School, which, by the way, was around the corner from Sullivan School. However, if Edward Pickering had attended Sullivan School, in all likelihood, he would have then matriculated at Dixwell School (and not at Boston Latin, which, of course, we know he did). As such, we can rule out Sullivan School, which leaves Phelps School, founded by another noted Boston schoolmaster—Franklin Phelps.

Phelps School. According to several accounts, Franklin Phelps was a kindhearted schoolmaster, a man of the old school, who spoke straight but stood no boy's nonsense. The school, located at the corner of Chestnut and Charles Streets, was as spare and direct as its master. It had a rough bare floor and crudely made desks and chairs, each one bearing the etchings of its former occupants. There were no pictures or other ornaments hanging from the white-plastered walls, except for a large map at the front of the room.

Other than this, there is little known about Franklin Phelps or Phelps School, other than the fact that Phelps lived on West Cedar Street, only two blocks from the school, which in turn was only five blocks away from Edward Pickering's family residence on Mount Vernon Street. Although a bachelor, Phelps did not live alone. He shared a residence with another well-respected schoolmaster—Francis Gardner—who at the time was headmaster of Boston Latin. In fact, it was Phelps' job to prepare boys for Boston Latin, and Gardner's job to prepare them for Harvard.

Did Edward Charles Pickering attend Phelps School? In lieu of any existing school records or additional student remembrances, perhaps the old real estate adage applies here: *location, location, location.* Of the five private schools within walking distance of Edward Pickering's home on Mount Vernon Street, Phelps School was the closest, a mere stone's throw away. More than that, Phelps School, like Sullivan School, was a feeder school. Unlike Sullivan School, however, which prepared students for Dixwell School, Phelps School prepared young scholars for Boston Latin. In all probability, Edward Pickering attended Phelps School between the ages of six and ten years old, before matriculating at Boston Latin, then under the firm direction of Master Phelps's roommate, Francis Gardner.

And it would have been a delightful walk, as we have already seen in our earlier tour of Edward Pickering's South Slope neighborhood. To get there, Edward would have walked to the corner of Mount Vernon and Willow Streets and turned left where Louisburg Square meets Mount Vernon Street. From this intersection, Edward would have walked the two short blocks that make up Willow Street. At Chestnut Street, Edward would have turned right and walked the last two blocks to the corner of Chestnut and Charles Streets, perhaps stopping to listen to an early morning rehearsal by members of the Harvard Music Association located at the corner of Cedar and Chestnut Streets before arriving at Master Phelps' school.

Of course, this is only conjecture. I have no solid evidence other than what I've presented here. Be that as it may, the point of this exercise is to provide a window into the educational opportunities available to boys of upper-class Beacon Hill families whose sole goal was to send their sons to Harvard. The first step in that process, of course, was to find a solid preparatory school for them. More than likely, at least for Edward and Charlotte Pickering, Phelps School served that purpose, but it was only the first step: their greater goal was young Edward's matriculation at Boston Latin.

CHAPTER 3 | *Schools and Schoolmasters*

> *In spite of all revolutions and all the pressure of business and all the powerful influences inclining America to live in contemptuous ignorance of the rest of the world, and especially of the past, the Latin School, supported by the people of Boston, has kept the embers of traditional learning alive, at which the humblest rush-light might be lighted; has kept the highway clear for every boy to the professions of theology, law, medicine, and teaching, and a window open to his mind from these times to all other times and from this place to all other places.[1]*
> —George Santayana

Edward and Charlotte Pickering had only the best intentions in mind when they enrolled their eldest son in one of Boston's most cherished institutions—the Latin Grammar School of Boston (a.k.a. Boston Latin). Their goal from the start was to see Edward graduate from Harvard, as had his father, as well as other Pickering men, before him. To realize that goal, Edward and Charlotte Pickering knew that their son would need the best preparation possible. To attain that goal they had three choices: a private Latin grammar school run by a respected schoolmaster, an elite boarding school in the surrounding countryside, or, the only viable public option, Boston Latin. They chose Boston Latin because of its storied reputation as one of the best college preparatory schools in the area. But it was also a pragmatic choice: although they lived in Beacon Hill's South Slope neighborhood among the wealthiest families in Boston, they were not wealthy in the same manner as established Boston Brahmin families, who could afford to send their sons (and, later, daughters) to an elite private boarding school such as Phillips Exeter Academy in Exeter, New Hampshire.

For a thorough look at the early days of Boston Latin, there is only one work worth consulting: Pauline Holmes' *A Tercentenary History of the Boston Public Latin School, 1635-1935*.[2] It is a ten-year labor of love published in 1935 for the school's tercentennial celebration. The work, completed in partial fulfillment of the Masters of Arts degree at Wellesley College, was completed in two phases. While the first phase consisted of collecting and arranging chronologically over fourteen hundred documents, the second phase required the distillation and organization of the content of those documents into chapters. Using the contents of Holmes' book, I've done my own distillation and organization into relevant sections for this chapter. In particular, I break the period from 1635 to 1856 into epochs based on the various school buildings that housed the school. The above bookend dates are not arbitrary: while 1635 marks the school's inaugural year, 1856 marks the year Edward Pickering matriculated at Boston Latin.

From the earliest days of the Puritan experience in America, the founding members of the Massachusetts Bay Colony adopted and maintained the principle that it is the right and duty of the government to provide for the instruction of its young. The idea was formalized in the school law of 1642, which ordered, for the first time in the English-speaking world, that all children be taught to read and write. The law sprang from the experience of its founding members, many of whom were graduates of Cambridge or Oxford University. Quite naturally, they brought with them the spirit of English culture, which included an emphasis on scholarship. One of those early transplants was Rev. John Cotton, who immigrated to the New World in 1633 from Boston, England, a major city in the northern province of Lincolnshire. Rev. Cotton had been a powerful preacher and a respected community leader in his hometown, having been appointed to the town's vicarage in 1612.

Although we don't have any definitive records identifying Rev. Cotton with the founding of the Latin Grammar School in his newly adopted country, the school was founded less than two years after Cotton immigrated to Boston. Not only do we know that the Latin Grammar School of Boston was modeled after its English equivalent—the Free Grammar School of Boston, England, sanctioned by Queen Mary in 1551—but we also know that Rev. Cotton was nominated to determine the fitness of an usher (a junior teacher) for the Free Grammar School as early as 1613. Given these details, and the fact that Cotton graduated from Trinity College, Cambridge, and was elected afterwards to a fellowship at Emmanuel College, we can assume that Rev. Cotton had both an experiential and an intellectual understanding of the English educational system, an understanding that he drew upon when he established the first free Latin grammar school in the New World.

The Latin Grammar School of Boston was not only connected to its English counterpart through its curriculum, which emphasized training in Greek and Latin and a thorough understanding of the classics, but also through its emphasis on the "free" or public nature of the institution. The "free schools" of England were often founded and endowed by individuals in towns where there was no church-related school and were open to children living in the parish or township.

We find, however, that the descriptor "free" has a couple of meanings. The name "free school" was applied not only to the Latin Grammar School of Boston, but also to various "reading" and "writing" schools established in Boston during the colonial period. In this sense, a "free school" meant a democratic public institution not restricted to any class of children. Or, as Pauline Holmes writes: "Secondary education was declared at once not to be the privilege of any aristocratic class; the governor's son and the poor man's son were, in this sense, both 'free' to enter the school."[3]

The schools were "free" in another sense as well. Since they did not charge tuition fees (only a modest entrance or "fire fee," which was often waived for the poorest students), they were free in an economic sense. Of course, they were not free to run, so from their inception, the city subsidized their existence through several sources: voluntary contributions, income from town property, general town taxation, personal gifts of land and money, miscellaneous fees and fines, and tuition fees of non-residents. Despite the double connotation of the word "free," the

Latin Grammar School was not as free—in the democratic sense—as it appeared. Since the school was an exam school (entrance could only be gained by passing a rigorous set of exams), in truth, the Latin Grammar School was open only to those who could pass the exams, and to pass the exams, one needed a tutor—and that cost money. So, in reality, the Latin Grammar School of Boston was open—or free—to boys who came from wealthy families that could afford a private tutor. As such, from its inception, Boston Latin became a "fitting" or preparatory school for economically advantaged young men bound for Harvard College.

When Boston Latin was established in 1635, one year prior to the founding of Harvard College, its physical presence was in the home of its schoolmaster. For the first few years of its existence, that meant the homes of Philemon Pormort, who was schoolmaster during Boston Latin's first three years; Daniel Maude, Pormort's assistant and master of the school from 1638 to 1643; and John Woodbridge, who held the position of schoolmaster from 1643 until 1645. It was during Woodbridge's tenure as schoolmaster that the first schoolhouse was built on the north side of School Street behind King's Chapel on part of the Chapel's burial ground.

Like all early histories, they are often one part fact, one part conjecture. This seems to be the case with the early history of the first schoolhouse on the north side of School Street. According to Holmes, an entry in Boston's *Town Records* on March 31, 1645, indicates that Thomas Scottow sold his house, yard, and garden, located on the north side of School Street, to the city of Boston. The house was a simple structure, as most buildings were at the time, with Master Woodbridge living in a portion of it while keeping school in the rest. Holmes goes on to note that between 1645 and 1652, another building was built on the property "betwixt the towne's house in which Mr. Woodmansey now liveth, and the town skoole house."[4]

Henry Jenks, on the other hand, who compiled one of the earliest histories of Boston Latin, doesn't mention the Scottow property at all, stating only that a schoolhouse was built in 1645 on the north side of School Street, behind King's Chapel, on part of the burial ground. The schoolhouse, according to a rendering included in both Holmes' and Jenks' books, is a simple two-story brick building, with a three-step stoop leading to a single wooden door that faces School Street. On either side of the door there is one window with a heavy stone lintel. Three windows of equal size adorn the front side of the second floor. What makes this entry even more mysterious is that both authors date the schoolhouse from 1635, which we know is incorrect as the school met in the home of Philemon Pormort, the school's first schoolmaster. Without an independent arbitrator, how are we to make sense of these discrepancies? As I said above, early histories are one part fact, one part speculation. So let's speculate a bit.

More than likely the Latin Grammar School of Boston held classes in the home of its first three schoolmasters—Pormort, Maude, and Woodbridge—at which point, around 1645, the city purchased Thomas Scottow's house and property on School Street, behind King's Chapel, using the house as both the schoolmaster's residence and a place of instruction. Then, sometime between 1645 and 1652, a new schoolhouse was built on the property, that schoolhouse possibly being the building alluded to earlier by Holmes that was built between Mr. Woodmansey's house and the town's school house.

If this were the case, Robert Woodmansey would have been responsible for its construction, as he served as schoolmaster from 1650 to 1667. Here again the fog of history thickens, since there is no mention of a schoolmaster between 1645 and 1650, which means that if a schoolhouse were built on the Scottow property while John Woodbridge was schoolmaster, it would have been built in 1645, the year the property was acquired by the city and the year Woodbridge resigned. If the schoolhouse were built while Robert Woodmansey was schoolmaster, then it would have been built between 1650 and 1652 (if we are to believe Holmes' account that a building was built on the Scottow property between 1645 and 1652). In any case, by 1652 a two-story brick building stood on the north side of School Street, on the Scottow property, and is considered Boston Latin's first schoolhouse.

During this year, and for the next fifteen years, Robert Woodmansey held the position of schoolmaster and taught in the building until he resigned in 1667, at which time noted physician and epic poet Benjamin Tompson took over as schoolmaster, but only for a few years. In 1671, members of the Boston School Committee hired Ezekiel Cheever to head the school, a position he held for thirty-seven years.

Born in London in 1614, Cheever immigrated to Boston in 1637 after studying at Emmanuel College, Cambridge. He settled in New Haven the following spring after his appointment as master of the town's Latin grammar school. He subsequently served as master of two other Latin grammar schools, each located within the Commonwealth of Massachusetts—Ipswich (1650-1661) and Charlestown (1661-1671)—before serving as Boston Latin's headmaster from 1671 to 1708. Ezekiel Cheever is credited with the authorship of several important books, including *Scripture Prophecies Explained* and an introductory Latin grammar book used at the school and popularly known as *Cheever's Accidence*.[5]

Ezekiel Cheever died in 1708, the first master of Boston Latin School to die while in office. At a memorial ceremony held in Cheever's honor many years later, Governor Thomas Hutchinson said this of the school's headmaster: "August 21st, this year, died Ezekiel Cheever, venerable not merely for his great age, 94, but for having been the schoolmaster of most of the principal gentlemen in Boston who were then upon the stage."[6] And, indeed, it was true. Other masters may have held the reins of power longer than Cheever, but none made a greater impact on the school, at least in its earliest days.

Under Cheever, Boston Latin saw several significant changes—to its physical structure, location, curriculum, and even its name. Regarding the latter, in 1698, Cheever argued to have the school's name—the Free Grammar School of Boston (also known as the Latin Grammar School of Boston)—changed to the more simple and elegant Boston Latin School. The name change didn't last long, however. In 1712, the city of Boston opened another Latin grammar school, which it named the North Grammar School. While the North Grammar School was open over the next seven decades, Boston Latin went by the name South Grammar School to distinguish it from its northern counterpart. However, after passage of the school reform laws of 1789, and closure of the North Grammar School, the South Grammar School reverted to the name by which we know it today: Cheever's more concise and elegant Boston Latin School (or, as I have already mentioned, more simply—and affectionately—Boston Latin).[7]

By the end of the seventeenth century, with the schoolhouse and master's residence over fifty years old, Boston's city council began to discuss two issues related to Boston Latin: the schoolmaster's residence and the need for a new schoolhouse. According to the city records, on March 11, 1700, the city council voted to consider three options regarding the schoolmaster's residence, the old Scottow home in which Master Cheever lived: repairing the house, replacing it with a new structure, or renting a house on another piece of property. Almost a year passed before the council voted to build a new house on the site of the old Scottow home for its schoolmaster. Two months after the vote, on May 3, 1701, Master Cheever took up temporary housing while the new residence was built.

Once the question of the schoolmaster's residence was settled, the city council took up the question of a new schoolhouse. The vote in favor of a new building was taken in early 1704, with a description of the plan contained in the *Selectmen's Minutes* for July 24, 1704. The entry indicates that John Barnerd was to be paid one hundred pounds to oversee construction of the two-story structure. Although all records point to the construction of a schoolhouse on the north side of School Street around 1704, none of the accounts, including that of Holmes and Jenks, include a rendering of the structure. Not only that, but the rendering often presented as the "first" schoolhouse dated 1635 has an uncanny resemblance to the schoolhouse built in 1704.

Could it be, then, that the "first" schoolhouse used exclusively for instruction really was built in 1704, and that all previous schoolhouses were simply renovations of and/or additions to pre-existing structures on the Scottow property, and used both as schoolhouse and schoolmaster's residence? There's a good chance we'll never know. What we do know is that the mists that shroud our current speculations begin to evaporate with the construction of the next schoolhouse.

IN MARCH OF 1748, the proprietors of King's Chapel petitioned the City of Boston for the right to use land in the rear of the church, land that was occupied by Boston Latin School. John Lovell, who had become master of the school in 1734 (succeeding Nathaniel Williams, who, in turn, succeeded Ezekiel Cheever), vehemently opposed the decision. Notwithstanding Lovell's objections, residents of Boston voted to move the school. The affirmative vote came with a stipulation, however: the church was required to build a new schoolhouse on land purchased from Richard Saltonstall, which was on the south side of School Street almost directly across the street from the old schoolhouse.

Construction of the low, one-story brick building took almost a year, but on May 8, 1749, the doors opened with Master Lovell standing in the doorway, beckoning the school's young scholars to enter. The school commanded the corner of School Street and Cook's Court (now Chapman Place). Although it was a smaller building, it was decidedly more substantial, and elegantly topped with a cupola that gave it the gravitas of a Latin grammar school. The schoolhouse remained open, with Lovell at its helm, for the next twenty-six years, until April 19, 1775, when the British retreated to Boston after the battles of Lexington and Concord. On that day, upon learning the news of the battles, Master Lovell abruptly announced: "War's begun. School's done."[8]

Lovell was a graduate of Boston Latin and Harvard College, but he was not a patriot. Lovell was a Loyalist, a staunch supporter of the British. He began his teaching duties at Boston Latin in 1729 as an usher, teaching courses to the lower classes. After receiving his master's degree from Harvard in 1731, Lovell accepted the position of schoolmaster three years later. His mastership spanned four decades, until that fateful day in 1775 when the British attacked. Lovell remained in Boston for the greater part of the year, fleeing to Halifax, Nova Scotia in March of 1776 with British troops under the leadership of General Gage. Lovell's son, James, accompanied them, but not because he was a staunch Loyalist. James was a patriot and, unfortunately, a prisoner of the British, captured after the battle of Bunker Hill. He remained in Halifax until November of 1776, at which time he was released in a prisoner exchange. Returning to Boston, he was immediately elected a delegate to the Continental Congress. As for his father, John Lovell never returned to Boston. He died in Halifax in 1778.

Boston Latin remained closed for over a year, from April 1775 to the summer of 1776. On June 5, 1776, the school opened its doors in the same schoolhouse at the same location on the south side of School Street, but with Samuel Hunt its new schoolmaster. The low-slung brick building was in constant use from its reopening until 1785, when it was temporarily closed for repairs. During its closure, Boston Latin moved to temporary quarters in Faneuil Hall. Several months later, it convened in an old barn on Cole Lane, and then moved to a building on Pemberton Hill, finally returning to the refurbished school on the south side of School Street, but not for long.

In 1812, seven years into William Bigelow's headmastership, fire ravaged the School Street building. So severe was the damage that the building had to be torn down and a new structure built. It was built in the Greek-Revival style, a popular architectural style of the time promoted by architect Charles Bulfinch of the Mount Vernon Proprietors (and currently serving as head of the Boston School Committee). It was an impressive three-story brick building with a solid granite front, somewhat of a cross between an imposing bank and an impenetrable fortress.

The doors opened in 1814 with Benjamin Gould at the helm. Gould, like many of his predecessors, was a Harvard graduate, completing his master's degree in 1817, three years after he was appointed schoolmaster. He was an innovative educator: under his leadership, the curriculum was extended from a four-year course adopted in 1789 to a five-year course. Most notably, however, Gould instituted the practice of public student declamation. He also issued the first student report cards sent directly to each scholar's parents.

Frederic Leverett succeeded Gould upon the latter's retirement in 1828 and remained schoolmaster for the next three years until Charles Dillaway took over the reins of power in 1831. Neither schoolmaster left a notable legacy. But that would change with the next appointment. In 1836, Epes Sargent Dixwell was appointed schoolmaster of Boston Latin and remained its head master for the next fifteen years. Like many of his predecessors, Dixwell was a graduate of both Boston Latin and Harvard College. He came with significant educational experience, having served both as an usher at the English High School and as sub-master at Boston Latin.

Unlike his immediate predecessors, Dixwell left a significant mark on the school. Most notably, he founded the Boston Latin School Association and made Benjamin Gould's dream of a school library a reality. He was totally dedicated to the school, especially the *idea* of the school: that it be a rigorous classical liberal arts education based on accurate scholarship. He himself was an excellent Latin scholar, translating English verses into Latin late into his life. Along with holding a high standard of scholarship, Dixwell held a high standard of moral behavior in all phases of his life. As a schoolmaster, that meant treating each scholar fairly, which he did almost to a fault. A strict disciplinarian with a dignified bearing, as we have seen from Henry Cabot Lodge's tribute earlier, Dixwell was not only a superb teacher, but also an apt manager of the institution. Before he resigned, due to the residency law passed by the city council in 1851, the strict, hard-working, and well-respected schoolmaster oversaw the school's relocation to Bedford Street.

An architectural drawing of the Bedford Street Schoolhouse reveals a three-story stone structure built in the Greek-Revival style with large vertical windows on each floor. The eye ascends quickly to the second floor windows, which are bordered by faux Greek columns, and then to the symmetrical cast-concrete frieze beneath the eave of the gently sloping roofline. Surrounded by a six-foot-tall wood-slat fence, the school looks—as we saw before with the second schoolhouse on the south side of School Street—like an impenetrable fortress.

By 1844, the year the Bedford Street schoolhouse opened, Boston Latin, which at the time went by the name South Grammar School, was only one among many public schools in the Boston public school system. A decade earlier, Boston saw the rise of Horace Mann as the Commonwealth's first Secretary of Education. It was Mann who pushed for more "common schools" to serve the needs of all young people in the Commonwealth of Massachusetts. At the secondary level, along with the South Grammar School and the North Grammar School, the city sponsored several "reading" and "writing" schools, intended for the general population, those more likely to wind up as manual workers. For those who wanted to be more than manual laborers, but were not bound for Harvard College, there was the English High School, established in 1821. When the South Grammar School moved into its new schoolhouse on Bedford Street, so did the English High School, sharing the massive structure for the next 37 years.

Although Master Dixwell guided Boston Latin School's relocation to Bedford Street with deftness, it was Dixwell's successor, Francis Gardner, who put his stamp on the Bedford Street location for the next quarter of a century. Like many of his predecessors, Gardner was a graduate of Boston Latin (1827) and Harvard (B.A. 1831, M.A. 1834). And, like other schoolmasters, Gardner rose through the ranks of the school, first as usher (1831-1836), then as sub-master (1836-1850), and finally as master (1851-1876). In a memorial address given by William Dimmock in honor of the deceased schoolmaster, Dimmock said this about Gardner:

> He had a certain grim humor and an odd quaintness of expression that were very effective in his dealing with the boys, and often very amusing as they were repeated and passed through the school. Conceit he was very apt to pierce, and the few pupils who carried away from the school any bitterness of feeling, were mainly those whom, after the

manner of Socrates, he chided for "thinking themselves to be something when they were nothing."[9]

Other tributes describe Gardner as fully dedicated to the school and to his charges. He exuded the ideals of classical education through every fiber in his body. Rugged and forthright, he made both friends and enemies. The latter was especially so toward the end of his tenure as headmaster, when he vehemently opposed—though unsuccessfully—the imposition of a "general culture" curriculum on the school by the Boston School Committee.

Gardner was a classicist at heart, unabashedly and incontrovertibly. One only has to read the series of Latin School textbooks he edited to understand this. When he died, in 1876, the first headmaster to die in office since Ezekiel Cheever, he was regarded as one of the most celebrated educators of mid-nineteenth-century Boston. In commemoration of the 250th anniversary of Boston Latin in 1885, Phillips Brooks, an Episcopal clergyman and longtime Rector of Boston's Trinity Church, wrote the following about Master Gardner:

> The character and work of Francis Gardner will furnish subjects of discussion as long as any men live who were his pupils, and perhaps long after the latest of his scholars shall have tottered to the grave. But certain things will always be clear regarding him, and will insure his perpetual remembrance, especially these two. His whole life was bound up in the school and its interests, and his originality and intensity of mind and nature exercised the strongest influence over the boys who passed under his charge.[10]

In 1881, five years after Gardner retired as the school's headmaster, Boston Latin moved to Warren Avenue and Dartmouth Street in South End. It remained there, sharing the building with the English High School until 1923, at which time Boston Latin moved to its current location, off Avenue Louis Pasteur in Fenway.

But these buildings don't concern us; for it was at the Bedford Street location that Edward Pickering received his Latin grammar school education under the sure hand of Francis Gardner. To get to the school, Edward would have walked east along Mount Vernon Street and then turned right onto Joy Street until he reached the Common. After crossing Beacon Street, he would cut diagonally across the northeastern tip of the Common before exiting the Common at West Street. Since West Street turned into Bedford Street after several blocks, Edward would simply follow West Street to Boston Latin's Bedford Street location. It was a path he would take for six years, from 1856 to 1862. And what did he learn at Boston Latin? To understand this, we'll have to retrace our steps a bit, going back to the founding of the Commonwealth of Massachusetts itself.

Scholars attending a Latin grammar school in the mid-seventeenth century studied for a purpose: they were the intended new leaders of the colonies, the next generation of teachers, ministers, lawyers, and doctors. To this extent, they studied what their fathers had studied in English grammar schools. They learned the rudiments of Latin and Greek, read and discussed the classics, practiced English grammar and composition, and studied mathematics until they were qualified to enter Harvard College. It was a curriculum based on the humanities, with a solid understanding of the English language, as well as a thorough exposure to Latin and Greek. In short, it was—as we have already seen—what Henry Dunster, the first president of Harvard College, expected of all entering scholars. The first printed record of subjects taught and textbooks used appeared in 1708 by Rev. Cotton Mather, who attended Boston Latin in the 1670s. In *Corderius Americanus,* the printed edition of a funeral sermon and elegy for schoolmaster Ezekiel Cheever, Rev. Mather recounts some of the authors he studied, which included Ovid, Cicero, Virgil, Homer, and Cato.

In 1712, Nathaniel Williams, the school's master from 1708-1734, gives an even fuller account of subjects studied in a letter sent to Nehemiah Hobart, then senior fellow of Harvard College. In the letter, Williams highlights what scholars studied throughout the seven-year course of study, emphasizing the importance of English grammar, orthography, and composition, Latin and Greek grammar, and the reading of various Latin and Greek classics. There was also an emphasis on translating English verses into Latin and on turning a theme into a declamation or public recitation before a general assembly of other students and teachers.[11]

Several other accounts exist during the first century of the school's existence, primarily firsthand accounts by former students. One in particular offers a clear picture of the serious and thoroughly rigorous approach to learning at the school. Rev. Jonathan Homer, who studied under Master Lovell between 1766 and 1773, recalls many of the textbooks studied and exercises completed, all fully resonant with what both Mather and Williams recount above, ending with this declaration: "I entered [Harvard College] at the age of fourteen years and three months, and was equal in Latin and Greek to the best in the senior class. Xenophon and Sallust were the only books used in college that I had not studied."[12] Clearly, the Latin Grammar School of Boston was designed as a college preparatory school; and, just as clearly, almost a century and a half after the school's founding, it was still fulfilling its stated mission.

At the Massachusetts Constitutional Convention held in 1779, John Adams spearheaded the drafting of the Constitution of the Commonwealth of Massachusetts, which was approved by voters on June 15, 1780. The document, still one of the oldest state constitutions in force today, stressed the importance of education for the young citizens of the Commonwealth, and prompted the school reform laws passed less than a decade later. On June 25, 1789, nine years after the constitution went into effect, the Massachusetts General Court approved a school reform proposal called the New System of Public Education. Although the reforms impacted education throughout the Commonwealth, they impacted Boston Latin quite significantly. The most important change was a reduction of the seven-year course of study, patterned after the English grammar school model, to four years. The entry age for the school was lowered as well, with all candidates required to be at least ten years of age upon entry instead of twelve years, with each candidate already well instructed in English grammar.

The reform laws also prompted the closing of the North Grammar School, leaving the South Grammar School—renamed Boston Latin School—the sole surviving Latin grammar school in the city, supplemented by three "reading" schools and three "writing" schools. It would be another thirty years before the English High School, a general curriculum secondary school with a vocational orientation, would open its doors. Like its seven-year predecessor, Boston Latin's four-year course of study leaned heavily on English and Latin grammar, rhetoric, and composition, as well as constant translations of English works into Latin and vice versa. At the heart of the curriculum were Ezekiel Cheever's popular *Accidence* and Cotton Mather's *Corderius Americanus*, as well as several other works pertaining to the study of Latin and Greek, including John Ward's *Latin Grammar* and *Latin Accidence*, John Garretson's *English Exercises for School-boys to Translate into Latin*, John Clarke's *Introduction to the Making of Latin*, and William King's *History of the Heathen Gods*. And, of course, there was the reading of various English, Latin, and Greek selections from sources such as Aesop, Ovid, Virgil, Horace, and Homer, along with a thorough study of Tully's *Epistles* and *Offices*.

A year after the New System of Public Education went into effect, causing significant changes to Boston Latin's curriculum, another momentous event occurred: Benjamin Franklin died in Philadelphia on April 17, 1790, at the age of 84. Franklin, a giant of an intellect who immersed himself in every social, political, and scientific interest of his age, was also a student at Boston Latin, but not a graduate, having dropped out after one year of study (a second year of study at George Brownell's writing school followed and that was it: no more public education for the gregarious polymath). Even so, Franklin bequeathed to Boston Latin one hundred pounds sterling, the interest of which was to be spent on silver medals—called the Franklin Medals—to be awarded each year to the school's best scholars.

The best scholars were identified not only by their consistently high marks from one year to the next, but also by a system of public speaking called declamation, one of the school's most time-honored traditions.

Each year, students were required to give an oration in English class at least three times during the year. Students were scored on a variety of aspects, including memorization, presentation, and enunciation. At the end of the school year, those who scored well were given the chance to declaim in front of a panel of alumni judges, ushers, schoolmasters, and special guests, with the winners garnering the coveted Franklin Medal.

Aside from a drop in the school's entrance age from twelve to ten years old and the reduction of the seven-year program to four years (and then increased to six years), Boston Latin saw very little change in its curriculum between the school reform laws of 1789 and the year Edward Pickering entered the school in 1856 at the age of ten. Harvard president Charles Eliot, who began his studies at Boston Latin in 1844, attested to the strength and continuity of the school's mission in an address he delivered in 1910 at Boston Latin's 275th anniversary:

> Sixty years ago, when I entered [Boston Latin], the subjects of instruction were Latin, Greek, mathematics, English composition and declamation, and the elements of Greek and Roman history. There was no formal instruction in the English language and literature, no modern language, no science, and no physical training, or military drill. In short, the subjects of instruction were what they had been for two hundred years.[13]

It was a curriculum that served the residents of Boston well. The curriculum, although dry as a bone, but rigorous and enduring, equipped only too well the young scholars whose aim it was to attend Harvard College. In this sense, Boston's first Latin grammar school has stayed true to its roots, remaining the nation's first free public college-preparatory school of the highest standard. This is the curriculum that not only equipped Edward Pickering to enter Harvard, but also to be successful at his studies.

Had Edward Pickering been born twenty years later, however, attending Boston Latin during the 1870s, he would have had another choice. To supplement the original classical curriculum, overseers of

Boston Latin, over the strong objections of headmaster Francis Gardner, voted to institute a general curriculum at the beginning of the 1870-71 school year. The general curriculum differed from the traditional Latin grammar school curriculum in its emphasis on a broader array of contemporary themes and subjects, which included Latin, Greek, French, German, American literature, English literature, ancient, medieval, and modern history, mathematics, geography, zoology, geology, botany, physics, astronomy, chemistry, and philosophy. However, given its breadth and scope, the curriculum proved somewhat impractical as a college-preparatory program. Furthermore, it mirrored too closely the curriculum of the English High School that shared the same building. Given these limitations, Boston Latin abandoned the general curriculum in 1876 and returned with renewed energy and dedication to the original Latin grammar school curriculum.

In retrospect, it's a shame that Edward Pickering didn't attend Boston Latin during the 1870s, when the school offered the general curriculum. Given his interest in science—physics in particular—he would have undoubtedly thrived. Unlike his younger brother, William Henry Pickering, Edward didn't excel at school; unlike his brother, he never won a Franklin Medal, or any other award. He was, in short, a rather ordinary student; well prepared, yes, but unenthusiastic about his future course of studies. Perhaps that is why future Harvard president Charles Eliot, who knew most of the senior class at Boston Latin personally, recommended that Edward not attend Harvard College, the four-year classical liberal arts program, as most of Edward's forebears had. Instead, Eliot suggested that Edward enter Harvard's Lawrence Scientific School, established in 1847 by Abbott Lawrence, a wealthy Massachusetts industrialist and entrepreneur. And that was exactly what Edward did. No one minded. No one objected. Not he, nor his parents. The Lawrence Scientific School seemed a good match for the intelligent and sensitive, but somewhat brooding, young scholar.

CHAPTER 4 | *The Leaven of Improvement*

> *Courses and professorships in science already existed in many colleges, some dating back to the previous century, but these were academic in character. The significant difference in the early years of the nineteenth century was that numerous voices were being raised, and a number of ventures proposed or begun, in behalf of both popular and practical or technical education, quite apart from the classical tradition.[1]* —Samuel Reznek

In the fall of 1862, one year after the outbreak of the American Civil War, Edward Pickering began his studies at Harvard's Lawrence Scientific School. The young scholar was sixteen years old, a year younger than the school itself, which was founded in 1847 with a gift from textile magnate Abbott Lawrence. At sixteen years old, Edward must have been wide-eyed at the prospect of studying science and technology, two subjects close to his heart and a far cry from the classical liberal arts curriculum he had been subjected to at Boston Latin.

But he would not have had the opportunity to study science at Harvard, or any other college of his day, had it not been for changing attitudes toward science since the colonies won their independence from Britain in the 1780s. The Revolutionary War, a grueling seven-year experience, served as a tremendous catalyst to the new country, a spark igniting the hearts and minds of its citizenry. The overarching theme in the early days of the post-Revolutionary War period—beyond the basic desire for freedom—was an overwhelming thirst for knowledge, a thirst driven by two forces: the specific desire to improve one's circumstance and a general desire for the social and economic progress of the new republic.

The intersection of these two forces was no more palpable in nineteenth-century America than in the burgeoning city of Boston, often referred to as the "Athens of America."[2] It was here that many of America's early educational institutions were envisioned, created, and tested. Not only did the greater Boston area host the country's first Latin grammar school and college, it was also the home of Horace Mann, the nation's first state secretary of education and one of the founders of the Common School Movement. It is no wonder that Boston became the hub of many other educational initiatives as well. Being the civic-minded town that it was, Boston saw the organization of a number of learned societies, technical institutes, and public lecture programs, each focused on the dissemination of knowledge, especially of a useful, practical kind.

One of the first groups to address the needs of unskilled young workers was the Massachusetts Charitable Mechanics Association, founded initially to stem the tide of apprentices fleeing their apprenticeships before their contract was up. The apprenticeship, which was an age-old European custom that provided master craftsmen with low-cost workers while training young boys in a trade that could sustain them throughout their adult life, worked well unless masters abused the power of their office or their apprentices—once skilled—chose to leave their appointment prematurely in order to seek work at journeymen's wages.

In Boston, as elsewhere, the problem of runaway apprentices was a nagging concern. In response to the problem several Bostonians, including Paul Revere who ran a copper-sheathing business, stepped up to address the issue. The result was the formation of the Massachusetts Charitable Mechanics Association in 1795. The group's annual report a century later outlined the persistent problem:

> At that time it was usual for master mechanics in every line of industry to keep apprentices, bound by agreement to serve them during their minority. These being practically trained in all the details of their respective trades would, in many instances, after attaining their majority, in turn become masters themselves, thus perpetuating the line of master mechanics personally and thoroughly skilled in their

respective callings. Such apprentices frequently became expert workmen before the expiration of their terms of service, and occasionally were dishonest enough to leave their masters and seek employment elsewhere, in order sooner to obtain journeymen's wages.[3]

For the first few years of its existence, the Association focused primarily on the apprentice problem, developing a certificate of completion that was given to each apprentice at the successful conclusion of his apprenticeship, a certificate he had to produce in order to be hired as a journeyman. That seemed to stem the tide of most premature departures; as such, the group turned its attention to providing social and educational opportunities for its members and their apprentices. As a result, within a few short years, the Massachusetts Charitable Mechanics Association hosted lectures, exhibitions, conferences, evening classes, as well as a library and a school—all focused on supporting tradesmen and their apprentices in the mechanic or practical arts. In other words, the group's answer to the apprenticeship problem—beyond a certificate of completion—was a bold, progressive plan of education and technical training.

One of the most beneficial and lasting outcomes of the Association was its library, which came about in 1820, when Boston merchant William Wood informed members that he was interested in donating several hundred books to the group as the nucleus of a lending library for use by members' apprentices. Within a few years, Wood's interest blossomed into a full-fledged library of over fifteen hundred volumes. However, the time and effort needed to maintain the library became so great that the group's members voted to hand the administration of the library over to the apprentices themselves. In 1828, they did just that, encouraging the apprentices to form their own organization—the Mechanic Apprentices Library Association. For the next several decades, members of the Mechanic Apprentices Library Association supervised the book-lending program, solicited new donations, monitored monthly subscriptions, purchased, processed, and shelved new books, and generally helped maintain and run the library.

Whereas the focus of the Massachusetts Charitable Mechanics Association was on the working-class adult, it did not ignore the public school system. Leaders of the group often had a close working relationship with members of the Boston School Committee, an institution that dated back to the mid-seventeenth century, with members often sharing leadership positions in both organizations. This was not out of the ordinary, rather another example of the far-reaching arm of any city's business community, a community interested in many things, but especially in producing skilled workers to fill the ranks of their manufacturing, mercantile, and banking endeavors.

Since existing forms of education in and around Boston—the town school, the private academy, the Latin grammar school, and the classically-oriented college—didn't address these needs sufficiently, members of the Massachusetts Charitable Mechanics Association took it upon themselves to develop a free, tax-supported secondary school with a strong vocational emphasis. Boston already had a free, tax-supported secondary school—The Latin Grammar School of Boston—but that school existed for a small percentage of the population. Since admission to Boston Latin depended upon a student's ability to pass a set of rigorous entrance exams, it presumed that a family could afford to send its sons to a private academy or to hire a tutor to prepare them for the exams. As such, the desks of Boston Latin were often filled with the sons (daughters would come later) of those who could most afford to prepare them for entry into the school.

What was needed was another secondary school experience, one that was also free and tax-supported, but open to the sons (again, daughters would come later) of the working class. A school where, along with grammar, rhetoric, English literature and composition, moral philosophy, history, civics, and geography, students could take courses in surveying, navigation, map reading, astronomy, optics, bookkeeping, even penmanship as they prepared for careers as accountants, surveyors, notaries, clerks, merchants, bankers, and the like.

In 1821, one year after Boston became incorporated as a city and almost two hundred years after the founding of Boston Latin, Boston's city council, with the backing and encouragement of the Massachusetts Charitable Mechanics Association, opened the country's first free public high school for working-class families. Originally named the English Classical School, the institution was renamed the English High School in 1824, when it moved from the corner of Derne and Temple Streets near the Massachusetts State House to the corner of Pinckney and Anderson Streets in Beacon Hill's North Slope neighborhood. Since the "classical" liberal arts curriculum of the Latin grammar school was all but absent, it seemed only right to delete the reference from the school's name.

The English High School was public to the extent that it admitted boys at least twelve years of age who had taken and passed the Boston Public Schools' basic literacy tests but were not interested in taking the more rigorous exams that the Boston Latin School required for entry. As such, the school attracted the sons of Boston's working class intent upon receiving a solid secondary school education, but not necessarily interested in pursuing additional studies at the collegiate level. From its inception, it would set the standard for public high schools in America, which is reflected in Rev. James Fraser's address to the British Parliament several decades after the school's founding. Referring to the English High School in Boston, Fraser writes:

> It is the one above all others that I visited in America which I should like the Commissioners to have seen at work, as I myself saw it at work on the tenth of June, the very type of a school for the middle classes of this country, managed in the most admirable spirit, and attended by just the sort of boys one would desire to see in such a school. Take it for all in all, and as accomplishing the end at which it professes to aim, the English High School at Boston struck me as the model school of the United States.[4]

THE FOUNDING OF BOSTON'S English High School was in many ways the expression of a larger project that would soon have America in its grip—the American Lyceum Movement—founded by farmer-turned-educator Josiah Holbrook, who articulated what others felt intuitively: that post-Revolutionary War America needed a system of education dedicated to the systematic diffusion of useful, scientific-based knowledge for working-class families. Drawing upon existing European models, especially the mechanic institutes popular in the British Isles, Holbrook envisioned a network of community-based groups that would meet on a regular basis in order to discuss, debate, and disseminate the latest technical and scientific knowledge. It was a response both to the increasing interest in practical or useful knowledge, but also to the lack of educational opportunity for those who had chosen—or were forced by circumstance—to enter the workforce at a young age rather than pursue professional studies.

Born into a wealthy farm family in Derby, Connecticut, and educated in New Haven under the influence of Benjamin Silliman, Yale's first professor of chemistry and natural history, Holbrook was the right man at the right time. Teacher, lecturer, science enthusiast, and visionary educator rolled into one, Holbrook believed fervently that education should be a lifelong experience, an experience forged communally through a network of town meetings or lyceums dedicated to improving the social, intellectual, and moral fabric of society. Holbrook envisioned a network of voluntary associations, local in nature, which would contribute toward the spread of learning, especially in the mechanical sciences, while fulfilling the general thirst for knowledge that was sweeping the country at the time.

The choice of the word "lyceum" was not accidental. Holbrook wanted to make a conscious connection to the ancient Greek experiment in democracy, bringing that reference forward to the democratic experiment fully underway in early nineteenth-century America. And what better way to support the fledgling democracy than by proposing a method of adult education that was democratic, non-hierarchical, and collaborative. Holbrook envisioned nothing less than a series of tight-knit community-based associations of working adults committed to self-improvement through mutual education.

Acutely aware of the fact that practical, scientific-based education was not being sufficiently addressed in town schools, Latin grammar schools, private academies, and four-year colleges, Holbrook outlined his vision in the October 1826 issue of the *American Journal of Education* in an anonymous letter to the editor that called for the formation of "associations for mutual instruction in the sciences, and in useful knowledge generally."[5] Within a month of the letter's publication, thirty or more factory workers and farmers heeded Holbrook's call and came together in Millbury, Massachusetts, to form Millbury Branch Number One of the American Lyceum.

Other groups existed prior to the formation of the Millbury lyceum with many of the same goals and interests, even with the word "lyceum" in their name (most notably, New York City's Lyceum of Natural History founded in 1817 and the Detroit Lyceum founded in 1818). The founding of the Millbury lyceum in 1826, however, marked the beginning of the lyceum movement as an American institution. Not only was Holbrook the inspirational genius behind the movement, but he was also its most dogged promoter, advancing the lyceum cause through a torrent of pamphlets, circulars, and instructional materials, all supported by a thunderstorm of newspaper reports chronicling his travels to out-of-the-way communities to promote the idea of community-based mutual education societies. He was, in essence, the Johnny Appleseed of the American Lyceum Movement, traveling from one town to the next spreading the gospel of mutual education societies.

More than any other educational structure, Holbrook believed that the town lyceum had the greatest chance of contributing toward individual improvement, which, in turn, would further the social and economic progress necessary for the American democratic experiment to succeed. As popular as the movement was (by 1939, more than four thousand lyceums had formed, mostly along the Eastern Seaboard and in the Midwest), the lyceum movement would deteriorate in both vision and influence through several discernable phases. In the earliest phase of the movement, between 1826 and 1850, the emphasis was clearly on mutual education societies that served the working class. This was the utopian vision forged out of the republican spirit sweeping the country that Holbrook so eloquently wrote about. From the start, the lyceum movement was dedicated to the spread of utilitarian knowledge among working-class adults. During this phase of the movement, blacksmiths, watchmakers, bookbinders, engine-builders, and other skilled laborers gathered in churches, schools, and town halls to listen to local speakers, to participate in heated debates, and to take part in mutual education classes.

As the Civil War approached, however, the lyceum became less a vehicle for mutual improvement and more a platform to air controversial topics of an overt political nature, such as slavery, voting rights, and westward expansion. Unfortunately, the rise of political content in a lyceum's lecture season usually signaled a decline in scientific content. But the war years demanded a public educational program designed to provoke as much as it was designed to inform.

After the Civil War, the lyceum movement saw its most profound change, becoming less a forum for the expression of controversy and more "a stage for comics, humorists, singers, and impersonators" with lectures, discussions, and debates slowly being replaced by theatrical performances, musical concerts, and historical impersonations.[6] Ironically, what contributed most to the decline of the lyceum movement was Holbrook's message itself: that education should be—indeed must be—a regular, lifelong experience. This idea, which was wholly consistent with the goals of the Common School Movement, contributed as much as anything to the slow demise of the community-based lyceum as the growth of government-mandated, tax-supported public education at the elementary and secondary levels gained nationwide support.

Despite the fact that the first town lyceum—Millbury Branch Number One—formed forty miles west of Boston, for all intents and purposes, the epicenter of the lyceum movement was Boston. It was a natural place for the movement to gain a foothold: Bostonians were curious, intelligent, prosperous, and civic-minded. Boston's middle class, in particular, who formed the backbone of the lyceum movement, were hard-working young men and women who saw the lyceum as a way to climb the city's social and economic ladder. Along with the influence of Josiah Holbrook, the lyceum movement in Boston gained traction through the efforts of Timothy Claxton, who immigrated to America in 1823 from Earsham, Norfolk County, England. After an extended tour of Russia to study workers' collectives, Claxton immigrated to the United States, settling thirty miles outside of Boston, in Methuen, Massachusetts, where he found work in a machine shop affiliated with a cotton factory. From an early age, Claxton showed great interest in the practical arts, with a related interest in mathematics and technical drawing, interests fostered as an apprentice to a tinsmith as a young adult.

While working in Methuen, Claxton revived a study group of men and women meeting to discuss books read to them by their leader, a minister at a local church. Influenced by the mechanics' institutes he observed in Glasgow and London, Claxton encouraged members to be more active in their study group, to discuss and debate topics, engage in mutual instruction, and to start a lending library. Like Holbrook, Claxton was a zealous promoter of the cause of mutual self-improvement, especially among tradesmen and their apprentices.

After moving to Boston in 1826, Claxton founded the Boston Mechanics' Institution. However, like many institutes and town lyceums, Claxton's Boston Mechanics' Institution was short-lived, becoming more of a staid lecture program than a dynamic society of mutual education. Undeterred, Claxton revived the idea in 1831, changing the name of the group to the Boston Mechanics' Lyceum, offering—in the original spirit of the lyceum movement—a series of lectures of scientific, technical, and mechanical interest to its audience. During the same year, Claxton published *Memoir of a Mechanic*, which was published in the popular Boston periodical *Young Mechanic*. In it, Claxton reflects on the initial motivation to write a memoir:

> It was thought by my friends that it would be useful towards the encouragement of young men who, like myself, begin the world poor, and deficient in education; and also in showing, in some degree, what may be done by industry, perseverance and economy of money and time, by those who are obliged to leave school at an early age, to attend to some mechanical occupation as a means of subsistence.[7]

Although primarily autobiographical in nature, Claxton's memoir also eloquently captures the spirit of the age: the acquisition of practical, useful knowledge of a scientific and technical nature was all but the rage in early nineteenth-century America, a century that began with Oliver Evans' invention of vapor-compression refrigeration and was quickly followed by Walter Hunt's lock-stitch sewing machine, Hiram Moore's combine harvester, William Otis's steam shovel, and Richard Hoe's rotary printing press. And this is to name only a few of the many inventions by American manufacturers, inventions that led ultimately to the Gilded Age of the 1880s.

In Boston, two organizations in particular stood out among the two dozen or so groups that met in the greater Boston area in the first half of the nineteenth century intent on discussing the rising interest in technical knowledge: the Boston Society for the Diffusion of Useful Knowledge and the Lowell Institute. Established in 1829, three years after Holbrook's initial call to arms, the Boston Society for the Diffusion of Useful Knowledge—also known as the Boston Lyceum—was modeled after a group with the same name that formed in London in the early 1820s. Its aim was to publish reading material of a scientific nature for the express purpose of self-education, especially for those who were unable to obtain a formal education. A civic group, the Boston Society for the Diffusion of Useful Knowledge was dedicated to popular education in the form of lectures, discussions, and classes of mutual instruction, beginning officially on August 13, 1829, with a series of classes, lectures, and debates. Like many lyceums and institutes that emerged during the 1820s, in the early days of its existence, the group met at various Boston locations, including the Boston Athenaeum, the Masonic Temple, Tremont Hall, and the Odeon or Federal Street Theater.

The initial meeting, called by newly elected Massachusetts State Senator Daniel Webster, featured Josiah Holbrook, whose name by the late 1820s was synonymous with the lyceum movement. Holbrook spoke before a gathering of some of Boston's most prominent citizens as Webster, the meeting's chairman, looked on. Within a few short weeks, the group formed the Boston Society for the Diffusion of Useful Knowledge. Though essentially a public forum operating outside the walls of higher education, a number of the society's founding members had strong ties to Harvard. These included Daniel Webster, Nathan Hale, Jacob Bigelow, William Ellery Channing, Edward Everett, and Edward Pickering's grandfather, John Pickering.[8]

As stimulating as the Society's lecture program was, it did not always satisfy the need for the systematic and rigorous study of the practical or mechanical arts, nor even introduce the audience to the basics of scientific thinking that was sweeping the country at the time. In "A Calendar of Lectures Presented by the Boston Society for the Diffusion of Useful Knowledge, 1829-1847," Helen Deese and Guy Woodall list some of the lectures given between 1829 and 1831, a list that underscores this point:

"The Biography of Franklin," Edward Everett, 1829
"The Causes of the Decline of the Turkish Empire," Francis Lieber, 1829
"The History of Civilization," Alexander Everett, 1830
"The Theory of Morals," Alonzo Potter, 1830
"On the Modern History of Massachusetts," James Austin, 1830
"On the Value of Human Knowledge," John Pierpont, 1830
"On Taxation and Revenue," John Gray, 1831
"On Laws of Property," Lemuel Shaw, 1831
"The Cemetery at Mount Auburn," Jacob Bigelow, 1831[9]

A new entry into the lyceum movement arrived in Boston in 1836, when industrialist John Lowell, Jr. bequeathed $250,000 for the creation of the eponymous Lowell Institute. A founding member of the Boston Society for the Diffusion of Useful Knowledge, Lowell established the Lowell Institute as a tribute to his wife and two children who had died within a two-year span. When Lowell fell ill several years later, he created a trust to endow an institute that would offer lectures and short courses on a variety of topics, including science and technology, liberal religion, literature, and the social sciences. Within a few short years, the Lowell Institute soon came to dominate Boston's thriving lyceum movement: by 1860, it had displaced or absorbed most of the competing lecture programs in the Boston area. In fact, between 1840-1860, starting with noted lecturer and Yale professor Benjamin Silliman, some sixty lecturers gave over one hundred presentations to several hundred thousand audience members.

Although the Lowell Institute had relationships with other cultural institutions, the direction and governance of the Institute remained firmly within the Lowell family circle, as decreed by Lowell's will that stipulated the Institute "always choose in preference to all others some male descendant of my grandfather, John Lowell, provided there be one who is competent to hold the office of trustee, and of the name of Lowell."[10] The overt nepotism paid off: the Lowell Institute outlived almost every other lyceum endeavor in Boston with the first trustee Lowell's cousin, John Amory Lowell, who administered the trust for more than forty years (who was succeeded in 1881 by his son, Augustus Lowell, who, in turn, was succeeded in 1900 by his son, Abbott Lawrence Lowell).

Along with a regular lecture program for the public, the Lowell Institute offered advanced courses of study on a variety of topics, many of which ultimately were spun off to affiliate institutions, namely the Museum of Fine Arts, the Franklin Institute of Technology, the Boston Society of Natural History, and the Massachusetts Institute of Technology. What was still missing, however—here and in every other endeavor—was the systematic introduction of scientific processes and rigorous training in the practical or mechanical arts. Since most antebellum colleges in the early nineteenth century didn't offer such training, another form of technical training was needed, training more focused and rigorous than what the apprenticeship, the private tutor, the Latin grammar school, the vocational high school, the town lyceum, or the mutual education training course could offer. The growing modernization of American industry demanded it. The group most affected by changes in technology was, of course, the merchant class, business leaders who, on a day-to-day basis, felt the pressing need to build factories, design machinery, transport raw materials, and, most importantly, hire skilled workers. If city fathers, well-heeled benefactors, or college presidents couldn't address this need, they would. And they did, by using their fortunes to create technical institutes to systematically train the next generation of skilled workers. But the business class wasn't the only group of individuals interested in training leaders in the rapidly developing fields of science and technology: the United States military was as well.

EVER SINCE OUR PURITAN forefathers set foot in the New World, education has been at the heart of the American experiment, driven by the need for an educated—or, at least, a literate—populace that could read and study scripture, but just as often propelled by the need for a skilled workforce that could keep pace with changing commercial pressures. By the beginning of the nineteenth century, the latter need—a skilled workforce—had become paramount due to the rapid acceleration of technological innovation; and just as innovation was sweeping through the commercial sector, so too was it sweeping through the education sector as evidenced by initiatives undertaken by the secondary grammar school, the private academy, the mutual education society, the technical institute's lecture hall, and even some classically-oriented college classrooms. What was on everyone's mind, except perhaps those thoroughly entrenched in colleges modeled after the English classical system, was how to teach the practical application of science to America's youth in order to meet the challenges of a new era. Surprisingly, the segment of society that did the most to embrace the study of science and technology as a core component in the education of America's youth at the dawn of the nineteenth century was not academia; it was the military.

Although George Washington envisioned it first, it was Thomas Jefferson who signed the Military Peace Establishment Act on March 16, 1802, authorizing Congress to establish a national military academy at West Point, a colonial fortification overlooking the Hudson River north of New York City.[11] Begun as an army outpost in 1778 below Gee's Point, a rocky outcrop that overlooked a sharp "S" turn in the river below, West Point enabled the Continental Army to defend the important waterway, preventing the British Royal Navy from sailing upriver in an attempt to divide the colonies. Briefly named Fort Arnold after Benedict Arnold (before the brilliant, but treasonous, commander tried to sell the fort to the British), the military academy was renamed twice: Fort Clinton, after Revolutionary War hero Major-General James Clinton, older brother of George Clinton, New York's first and longest-serving governor; and then, as we know it today, West Point, after its strategic geographic—and historic—location overlooking the Hudson River at a critical bend in the river.

Although the military academy at West Point floundered under inept leadership for the first decade and a half of its existence, operating more like a social club than a strict military academy, it found its bearings when President Monroe appointed Sylvanus Thayer as its superintendent in 1817. A descendant of a large and extended family from Braintree, Massachusetts, Thayer brought to the job several assets: first of all, he was valedictorian of his graduating class at Dartmouth College; next, he was a graduate of the Class of 1808 at West Point and a commissioned officer in the Corps of Engineers; and, finally, he had just completed a Grand Tour of Napoleonic France, studying at the famous École Polytechnique in Paris, a trip that was sponsored by the Madison administration, which provided funds for Thayer and his traveling companion William McRee to gather almanacs, maps, area surveys, and books on military tactics and history, mathematics, and engineering for use at West Point. It was widely thought, especially by Jefferson, who had served as U.S. Minister to France between 1784 and 1789, that if West Point was to succeed in educating military officers of the highest caliber, then imitating the French system was the most logical choice to follow.

Thayer and McRee left Boston aboard the United States frigate *Congress* on June 10, 1815, arriving in France in time to witness the occupation and sacking of Napoleonic Paris by England and its allies. For the next two years, the two commissioned officers visited one fortification after another, including Brest and Cherbourgh in western France. But it was in Paris that they did their most intensive work, especially at the École Polytechnique. Known for its emphasis on applied science, with engineering its foremost course of study, the École Polytechnique was one of the most prestigious of the French "grandes écoles." Established in 1794 by Gaspard Monge and Lazare Carnot at the time of the National Convention, the École Polytechnique quickly played a central role in the unfolding drama of the French Revolution, serving as a military academy guided by the Napoleonic motto *Pour la Patrie, les Sciences et la Gloire*.[12]

Even before Napoleon, who was trained as a military engineer and showed formidable mathematical skills, added his stamp of approval to the École Polytechnique, France was known in Europe and abroad as an advanced center of applied science and engineering. Without French assistance, the Continental Army couldn't have defeated Cornwallis at Yorktown, who was in the process of building a deep-harbor port at the mouth of the Chesapeake Bay in the fall of 1781. Along with French battleships that prevented Cornwallis from escaping into the Atlantic, and more than ten thousand French troops that prevented him from escaping by land, the French supplied a corps of trained army engineers and artillerists to support Washington's seven-thousand-man army. In short, not only did the American-French allied forces outnumber Cornwallis, but Washington and his French allied commanders outwitted him as well. It was precisely the role that French-educated military commanders and engineers played during the American War of Independence that led many to call for the establishment of a national military academy with engineering its central course of study.

Little did Sylvanus Thayer know that, upon his return from France in the summer of 1817, he—and not McRee, which Thayer had assumed—would be appointed superintendent of the military academy. His orders, given to him by interim Secretary of War George Graham almost immediately after he stepped onto American soil, were to proceed to West Point and to take over as superintendent of the military academy, which included the education, supervision, and commissioning of eligible cadets into the Corps of Engineers. It would not be an easy task since the academy, led by West Point graduate Alden Partridge, ran on a system of favoritism and patronage. But Thayer was more than prepared for the task: in his first year in office, he instituted the "Thayer Triangle," which equated military leadership with the cultivation of honor, discipline, and education.[13]

With Thayer's system, out the window went all forms of favoritism. Also out the window went the cadets' extended summer vacation, replaced by a summer encampment on the broad plain overlooking the Hudson River. Most of Thayer's work, however, had to do with the school's curriculum. Rejecting the traditional liberal arts curriculum of the colonial college based on the English classical model, Thayer envisioned a "scientific school" modeled after France's École Polytechnique that blended the military and engineering traditions of Napoleonic France, placing a premium on the education of competent engineers as a crucial asset for military readiness.

In addition to the above changes, Thayer also changed the cadet training program from an unspecified time period (anywhere from several months to several years) to a graduated four-year program that divided cadets into four tiers or ranks, equivalent to today's freshman, sophomore, junior, and senior class standing. In addition to this, Thayer set up a merit-based system that required daily recitations and end-of-year oral examinations; in fact, every aspect of a cadet's performance was evaluated and ranked as objectively as possible in an attempt to avoid the previous superintendent's system of favoritism.

Not only did Thayer's merit system, which ranked each cadet at the end of an academic term by class, determine what he would study the following year, but it was also the final adjudicator of which branch of the army the cadet would enter upon graduation and commissioning, with only the top cadets in their class entering the Corps of Engineers. Thayer did not do this alone: he listened to and supported his academic faculty; he established an academic advisory board to bring in outside opinions concerning the engineering curriculum; and he created a board of visitors comprised of distinguished individuals nominated annually by the Secretary of War who were invited to inspect the operation of the academy and to participate in the end-of-year examinations.

As Thayer's model began to take hold at West Point, the national government looked to the officers of the Corps of Engineers not only to defend the nation, but also to supervise the expansion of its infrastructure. Officers from the Corps of Engineers became involved in the construction of bridges, highways, and railroads. They were instrumental in developing the nation's waterways and, later, responsible for flood control projects. They supervised harbor construction, built jetties and piers, extended seawalls, and erected lighthouses. They organized survey teams to help map the western states and territories. In short, officers commissioned into the Corps of Engineers were responsible for building the infrastructure that allowed the country to grow from a weak agricultural society to one of the strongest industrial nations on earth. Summarizing the impact of West Point graduates on civilian projects in the early nineteenth century, Lawrence Grayson writes, "Of the engineering graduates engaged in public works before 1840, a sizable fraction were West Point graduates, and at least 30 percent of them served as chief engineers of important projects on railways, canals, docks, wharves, roads, and other nonmilitary activities."[14]

They had to: prior to 1825, West Point was the only institution in America offering a structured program in civil engineering. By the time Thayer resigned as superintendent of West Point in 1833, due to conflicts with the Jackson administration over some of his disciplinary actions and cadet expulsions, only a handful of institutions of higher education, offered courses in civil engineering.[15] One of those institutions was Rensselaer Polytechnic Institute, located a hundred miles north of West Point in Troy, New York, where the Mohawk River empties into the Hudson.

Stephen Van Rensselaer was not an educator; he was an extremely wealthy, politically-connected New Yorker who descended from a long line of Dutch immigrants who owned large tracts of land in the Upper Hudson River Valley. An average student, Van Rensselaer attended the College of New Jersey in Princeton and later Harvard. A resident of New York, he served in both houses of the New York State Legislature, held the office of Lieutenant Governor, and represented his upstate New York district in the U.S. House of Representatives. These were only some of the leadership positions he held. He was president of the commission that built the Erie Canal, president of the New York State Board of Agriculture, and chancellor of the New York Board of Regents. On a local and civic level, Van Rensselaer held executive positions in the Albany Institute of Art and History, the Albany Academy, the Mohawk and Hudson Railroad Company, and the New York State Lyceum Association. In short, throughout his life Stephen Van Rensselaer lived up to the expectations of someone of high social and economic standing, donating both his time and money to public improvement projects. It is no wonder then that on January 5, 1825, Rensselaer opened the doors to a private institute—Rensselaer School—located in Troy, New York, in Old Bank Place, a handsome Federal structure built in 1801 as the main office of Farmers' Bank.

To run the school, Rensselaer turned to Amos Eaton, a lawyer, geologist, botanist, and educational pioneer. As the school's senior professor and later its sole fiscal agent, Eaton was charged with converting Old Bank Place into a school and dormitory for full-time students, as well as living quarters for his family. Along with the residential facilities, Eaton outfitted the building with a suite of classrooms, a library, an assay room, a chemical laboratory, a natural history exhibit room, a laboratory to conduct mechanical experiments, and an observatory, albeit small, for the study of astronomy.

From the start, it was an experiment in education, reflecting the many facets of its headmaster, whose unabated curiosity and zeal for science education was nothing short of infectious. Along with the school's unique subject matter (unique for its time, that is: chemistry, botany, natural history, physics, mathematics, and astronomy), Eaton brought to the venture innovative teaching practices, which included individualized instruction, a student-centered cooperative classroom, easy access to instructors and teaching assistants, and a self-monitoring honor code. Eaton was also a tireless writer who conducted weekly correspondence with major experts in various fields of science, while writing and publishing textbooks, brochures, and bulletins—much like his contemporary Josiah Holbrook—in order to spread the gospel of science education far and wide.

For the first ten years of the school's existence, Eaton kept his focus on natural history, even offering an extended summer trip along the Erie Canal to study the flora and fauna of the waterway's ecosystem. By 1835, however, Eaton's interests had expanded beyond natural history and Rensselaer Institute (Eaton dropped "School" in favor of "Institute" several years after his arrival) began offering engineering courses. By 1837, the school had two distinct departments or "schools" of study: a Natural Science Department and an Engineering Corps. As the demand for engineers surged in the early nineteenth century, the engineering department became the central focus of the school.

Rensselaer was not the first civilian institute to offer an engineering program in early nineteenth-century America, however. That distinction goes to a military school founded in 1819 by Alden Partridge,

Sylvanus Thayer's nemesis at West Point. After Thayer replaced Partridge in 1817, Partridge established the American Literary, Scientific, and Military Academy in Norwich, Vermont. Closely imitating the West Point model, Partridge's school offered several courses in civil engineering, including courses on the construction of roads, locks, canals, and bridges. In 1834, Partridge changed the name of the American Literary, Scientific, and Military Academy to Norwich University. Although Norwich awarded its first engineering degree in 1837, it was Rensselaer Institute, two years earlier, that awarded the nation's first engineering degree by a civilian institution.

After the death of Rensselaer, the school's patron, in 1839, and Eaton three years later, Rensselaer Institute went through a trying time. The loss of Eaton, in particular, was a blow to the institution, since it was through the sheer force of his personality and the strength of his intellect that the school thrived. It floundered for the next four years under the lackluster leadership of George Cook, who was named senior professor and fiscal agent of the Institute on May 10, 1842, the day Eaton died. After Cook resigned four years later, the Board of Trustees named Benjamin Greene, a professor of mathematics at Washington College in Chestertown, Maryland, to head the school. It was a good choice; the young, energetic professor, who was a recent graduate of Rensselaer, had ambitious plans for the school.

Greene was born in Lebanon, New Hampshire, in 1817, the oldest of ten children in a farm family originally from Rhode Island. But he did not take up farming, as the oldest son of a farmer would have been expected to; instead, he became interested in the mechanic or industrial arts, taking various jobs at local ironworks and foundries. In 1841, while working at Foresdale Iron Works in Brandon, Vermont, Greene wrote to Amos Eaton, inquiring about Rensselaer's engineering program. Before the year was out, Greene had quit his job at Foresdale and enrolled at Rensselaer, signing up for both the natural science and civil engineering programs. He graduated in 1842, several months after Eaton died, but retained his connection to Rensselaer through correspondence with its new senior professor, George Cook. When Cook resigned at the end of 1846, Rensselaer's Board of Trustees chose Greene to succeed him.

From the outset, Greene exerted his will over the school, naming himself both senior professor and director (though the director's title didn't become official until 1850, when the state legislature voted to create it). A year later, Greene changed the name of the school from Rensselaer Institute to Rensselaer Polytechnic Institute (though, again, the school's name wasn't officially changed by the state legislature until 1861). Greene was always looking ahead, even if his ideas didn't reach fruition. Nevertheless, like Josiah Holbrook, he was the right man for the right job, his appointment coming as the country was awakening to the need of scientific and technological education. And, like Eaton before him, Greene was passionate about science education, but his vision was far more expansive than Eaton's: Greene sought to turn Rensselaer into a "true polytechnic," one that offered scientific and technical training suitable to the requirements of a rapidly emerging industrial society.

Although keen to modernize the school, Greene continued many of his predecessor's practices: mornings were still devoted to lectures and student recitations with afternoons reserved for practical activities in the laboratory or the field; and summers were filled with specimen-collecting field trips. But Greene's goal of making Rensselaer a full-fledged polytechnic school inevitably meant change. To this extent, Greene added new academic ranks and titles; he expanded the number of programs and degrees; he raised tuition to better meet the needs of the institution; but, most importantly, he oversaw the conversion of Rensselaer from a one-year degree-granting school, originally focused on the study of natural history, into a multi-year polytechnic institution specializing in engineering.

Although training engineers was its prime purpose, by the mid-1850s Rensselaer had a variety of programs that led to completion degrees, including two-year programs in civil engineering, topographical engineering, and general science. Sandwiched between a yearlong preparatory program and a culminating yearlong land-surveying course, the entire program marked the first four-year curriculum in engineering education in the United States. Over the next few years, Greene continued to refine his ideas, calling continuously for a true polytechnic school in the best of the French tradition. It would include two programs of study: a General School and a Technical School (with the latter subdivided into a General Studies program and a Special Studies program). It was a grand vision, echoing the breadth of France's École Polytechnique's curriculum, but Greene's vision turned out to be more grandiose than grand since his comprehensive plan for transforming Rensselaer into a polytechnic institute never fully materialized, as funds necessary for realizing his dream were never completely forthcoming.

Greene's continual push to expand Rensselaer beyond its original vision was both a product of Greene's ambition and an appreciation—perhaps "fear" is the better word—of the growing competition threatening the uniqueness of Rensselaer from institutions like Union College in Schenectady, New York, the Polytechnic College of the State of Pennsylvania, the Brooklyn Collegiate and Polytechnic Institute, and Cooper Union for the Advancement of Science and Art in New York City. By 1859, the year Greene resigned over disagreements with the Board of Trustees, Rensselaer no longer had a solitary, central, and secure position in the arena of scientific and technical education. On the contrary, as Samuel Rezneck observes, "it was threatened with submergence in the tidal wave of new projects and foundations of schools and courses in engineering and science spreading throughout the country."[16]

The "tidal wave" gained momentum from several experiments in science education in the early decades of the nineteenth century. Two of them we've already read about: the military school and the private polytechnic institute. These experiments, along with town lyceum associations and other science-oriented civic organizations, addressed the need for education in the applied sciences from outside the walls of academia. Within the walls of academia, similar experiments were underway, but they were painfully slow, since they had to overcome resistance from faculty at tradition-bound antebellum colleges. By the early nineteenth century, however, change was in the air. But it did not come all at once, and the context in which it came was complex with at least three major issues facing the small classically-oriented college: church versus state control in higher education; the classical curriculum versus the elective system; and the unified college versus the multi-faceted university. Each of these issues contributed to succeeding waves of change in all corners of higher education—public and private, large and small. In the next chapter, we'll look at how these issues impacted one college and the changes they brought.

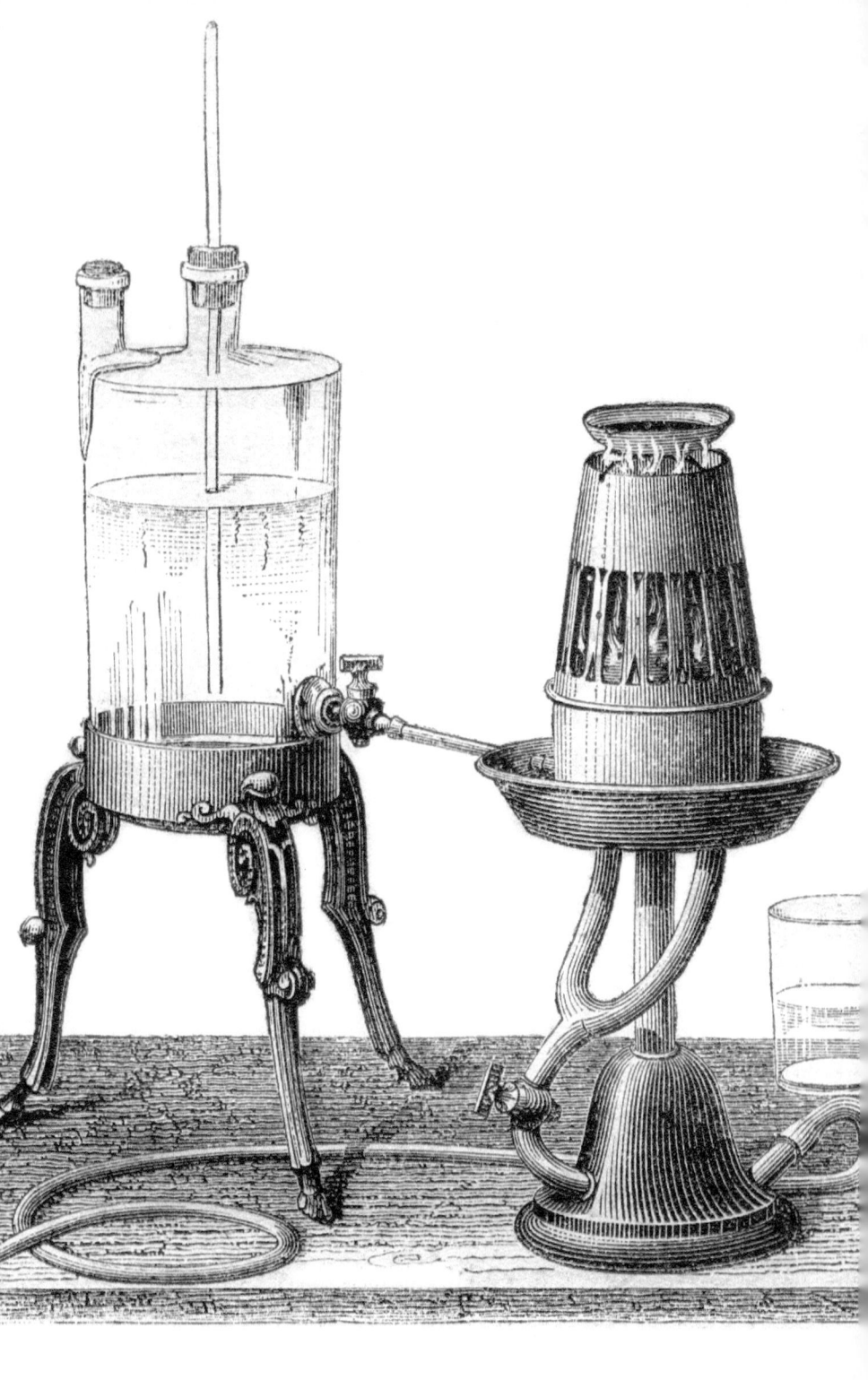

CHAPTER 5 | *"Old Sheff"*

> *Show me the way to Old Yale, boys,*
> *Show me the way to Old Yale;*
> *When head and hand grow weary,*
> *And the fountain of life doth fail;*
> *Show me the way to Old Yale, boys,*
> *Show me the way to Old Yale;*
> *—We have fought the fight, for truth and right—*
> *We have lived—let us die at Yale.*[1]
> *—Alfred E. Walker*

The traditional undergraduate curriculum at most early nineteenth-century colleges rested on the English classical liberal arts model, which had as its origins the medieval university curriculum based on the seven liberal arts, of which there were two parts. The first part, known as the *trivium*, was a systematic method of teaching critical thinking used to obtain factual certainty from information perceived by the senses. Concerned with the inner workings of the mind, the *trivium* was an input-process-output system that involved the "verbal arts" of grammar (input), logic (process), and rhetoric (output). On the other hand, the *quadrivium*, considered the higher of the two divisions, dealt with the secrets of nature through an exploration of the "number arts," i.e., arithmetic (number in the abstract), geometry (number in space), music (number in time), and astronomy (number in time and space). Both roads prepared the young scholar for the serious study of philosophy and theology. What did this mean on a practical basis? In *The American College and University: A History*, Frederick Rudolph outlines a typical four-year course of study found at most colleges during the colonial and early antebellum periods:

> During the first year Latin, Greek, logic, Hebrew, and rhetoric were the staples of the curriculum. During the second year logic, Greek, and Hebrew were continued, and a beginning was made on natural philosophy, which some centuries later would be called physics. In the third year there was added to natural philosophy mental philosophy or metaphysics and moral philosophy, a veritable grab bag of subject matter out of which one day would come economics, ethics, political science, and sociology. The fourth year provided review in Latin, Greek, logic, and natural philosophy. A modest beginning in mathematics was the only new departure of the senior year.[2]

This was the university model that immigrated to the New World with the first English colonists, forming the curricula basis for early colonial colleges whose interests were primarily religious (the education of future ministers) and professional (the training of doctors and lawyers). The two realms were not exclusive: both coalesced around the need to instill in students—that is, white male students from the upper classes—proper conduct, character development, and civic virtue. In other words, as Thomas Denham observes, paraphrasing Frederick Rudolph's work: "Imported from Europe, the curricula of the colonial colleges were creatures of both the Reformation and the Renaissance. They valued both the sectarian and the humanistic ideals of classical scholarship with the goal of creating learned clergymen, as well as gentlemen and scholars."[3]

After the American Revolution, however, the traditional four-year course of study began to see a subtle shift away from the medieval classical model that emphasized the moral and religious toward a European model influenced by the rise of scientific empiricism that emphasized the practical and scientific. By the end of the eighteenth century, new professorships began to appear in colleges that reflected this emphasis, especially in the field of natural history. As a result, a seismic shift was beginning to occur with many educators ready to throw out the traditional curriculum in order to usher in the new "modern" curriculum. Benjamin Rush, a renown physician from Philadelphia who served as Surgeon General of the Continental Army and later became a professor of chemistry, medical theory, and clinical practice at the University of Pennsylvania, echoed this sentiment in this memorable, albeit biting, observation: "To spend four or five years in learning two dead languages, is to turn our backs on a gold mine, in order to amuse ourselves catching butterflies."[4]

The two "dead languages" to which Rush refers, of course, are Latin and Greek, which by the second decade of the nineteenth century, had come under attack as useless and irrelevant. The growing attack hit a headwind in 1828, however, when the president and faculty of Yale College wrote a response to a suggestion put forth the previous year by trustees of Yale Corporation, urging the faculty to drop Latin and Greek from the academic or classical curriculum. The document, known as the Yale Report, was also a response to a more general and increasingly disturbing trend toward "a more fragmented, varied, vocational and specialized curriculum,"[5] which pitted advocates of the traditional liberal arts curriculum against those who favored a more progressive, practical curriculum, one that could respond more nimbly to the rapidly-evolving technological needs facing the new republic. As we have seen previously, exciting developments were beginning to stir outside the walls of academia: through the spread of town lyceums, the proliferation of mechanic associations, the rise of private technical institutes, and the efforts at several military academies. Summarizing the arguments of those advocating for change, Julie Kern writes:

> The growing concern about the question of curriculum in higher education was in large part due to the industrial revolution and the increase in agriculture in the period of the nineteenth century. Advocates for change believed that college should prepare a man for living, may it be through banking, farming, or industry. The curriculum should offer vocational education. Under this plan for reform, students would be able to pursue a specific plan of study so that he could learn a trade and become a positive contributor to the community in which he lived. Advocates for change

believed that curriculum should be relevant to commerce, industry, and agriculture.[6]

Although the tension between the traditional liberal arts curriculum and the newly emerging science-based curriculum didn't come to a head until the early nineteenth century, proposed changes to the traditional curriculum were seen as early as the middle of the eighteenth century. In 1754, Samuel Johnson of King's College (later Columbia University), endorsing Benjamin Franklin's philosophy of education, announced that King's College would emphasize surveying, navigation, geography, history, and natural philosophy, culminating in the knowledge "of Every Thing *useful* for the Comfort, Convenience and Elegance of Life...."[7] This was followed in 1756 by William Smith's proposal for a three-year course of study at the College of Philadelphia (later the University of Pennsylvania) that placed a full one-third of the curriculum's emphasis on the study of science and the mechanic arts, and as such became "the first systematic course in America not deriving from the medieval tradition nor intending to serve a religious purpose."[8]

At Yale, however, the river was about to reverse course: although the president and faculty of Yale College acknowledged the changing social, political, and economic dynamics of early nineteenth-century America, they argued that the best way to address those changing dynamics was through the retention of the traditional classically-oriented liberal arts curriculum. According to the authors of the report, represented primarily by President Jeremiah Day and James Kingsley, Professor of Hebrew, Greek, and Latin, the crux of the question was simple: What is the appropriate aim or object of the collegiate experience? According to the report's authors, the object of the experience should be nothing less than a superior and thorough education, one in which the ground work is "broad, and deep, and solid."[9]

The report comprised two major sections: the first part, written by President Day, was a general discussion of the nature of liberal arts education; the second part, written by Prof. Kingsley, was a more narrow argument for the retention of Greek and Latin literature in the college curriculum. It's the first part that concerns us the most, and begins with President Day staking out rather clearly—even passionately—the faculty's position:

> The two great points to be gained in intellectual culture are the *discipline* and the *furniture* of the mind; expanding its powers, and storing it with knowledge. The former of these is, perhaps, the more important of the two. A commanding object, therefore, in a collegiate course, should be, to call into daily and vigorous exercise the faculties of the student. Those branches of study should be prescribed, and those modes of instruction adopted, which are best calculated to teach the art of fixing the attention, directing the train of thought, analyzing a subject proposed for investigation; following, with accurate discrimination, the course of argument; balancing nicely the evidence presented to the judgment; awakening, elevating, and controlling the imagination; arranging, with skill, the treasures which memory gathers; rousing and guiding the powers of genius. All this is not to be effected by a light and hasty course of study; by reading a few books, hearing a few lectures, and spending some months at a literary institution. The habits of thinking are to be formed by long continued and close application. The mines of science must be penetrated far below the surface, before they will disclose their treasures. If a dexterous performance of the manual operations, in many of the mechanical arts, requires an apprenticeship, with diligent attention for years; much more does the training of the powers of the mind demand vigorous, and steady, and systematic effort.[10]

The excerpt is revealing on a number of accounts. First of all, it is divided into two parts: while the first part lays out the argument for the traditional liberal arts curriculum, its efficacy in establishing the "discipline" and "furniture" of the mind, the second part, through the effective use of metaphors of a scientific or technical nature, denigrates any type of manual arts training, which the author views as "light and hasty." What is called for, stated later in the report, is a balance between the different branches of literature and science, as to form in the student "a proper *balance* of character."[11] Indeed, reflecting the early educational aim of most colonial colleges, it is moral character with its overtones of religious study and devotion that is of utmost importance in the instruction of the young mind. To this end, the study of science is useful, but only—like the study of grammar, rhetoric, and logic—as it contributes to disciplining the mind and to expanding its contents or "furniture."

When it comes to disciplining the mind, let's look a little closer at how the liberal arts curriculum strengthens the mind's higher-order analytic powers. Here again we turn to the report itself where President Day lays out, subject by subject, the benefits to the student of the various branches of the liberal arts curriculum:

> From the pure mathematics, he learns the art of demonstrative reasoning. In attending to the physical sciences, he becomes familiar with facts, with the process of induction, and the varieties of probable evidence. In ancient literature, he finds some of the most finished models of taste. By English reading, he learns the powers of the language in which he is to speak and write. By logic and mental philosophy, he is taught the art of thinking; by rhetoric and oratory, the art of speaking. By frequent exercise on written composition, he acquires copiousness and accuracy of expression. By extemporaneous discussion, he becomes prompt and fluent, and animated.[12]

Day's argument rises to a crescendo with the following passage:
> The most important thing that a college experience can give a young person is the ability to think for themselves, to invoke the "resources of the mind." If a person cannot think for himself, does not have the discipline of mind to do so, which training in the liberal arts provides, then "the whole apparatus of libraries, and instruments, and specimens, and lectures, and teachers, will be insufficient to secure distinguished excellence."[13]

In other words, President Day concludes, the aim or object of a four-year course of study is not to finish the student's education, but to lay the foundation for advanced study in whatever profession or vocation the student so desires: "Our object is not to teach that which is peculiar to any one of the professions; but to lay the foundation which is common to them all."[14] There are two forces at work beneath the surface of this statement: on the one hand, an unapologetic embrace of the traditional liberal arts curriculum; on the other hand, a complete disdain for manual arts education at the undergraduate level, since such training is aimed at the trades and not at the esteemed professions of theology, law, and medicine. In referring to the manual arts, whether mercantile, mechanical, or agricultural, President Day writes:

> These can never be effectually learned except in the very circumstances in which they are to be practiced. The young merchant must be trained in the counting room, the mechanic, in the workshop, the farmer, in the field. But we have, on our premises, no experimental farm or retail shop; no cotton or iron manufactory; no hatter's, or silver-smith's, or coach-maker's establishment. For what purpose, then, it will be asked, are young men who are destined to these occupations, ever sent to college? They should not be sent, as we think, with an expectation of *finishing* their education at the college; but with a view of laying a thorough foundation in the principles of science, preparatory to the study of the practical arts.[15]

It is ironic that of all institutions of higher learning, Yale should draw the proverbial line in the sand regarding the need to preserve the traditional liberal arts curriculum; ironic because on the academic faculty at the time the report was written sat none other than Benjamin Silliman, regarded by many as the Dean of American Science. If anyone hastened the demise of the prescribed undergraduate liberal arts experience, it was Silliman, though he would not admit it, as he saw no conflict between the traditional undergraduate curriculum and the teaching of applied science to qualified "special" or advanced students (students, in other words, who had successfully navigated the prescribed undergraduate liberal arts curriculum prior to working in his chemistry laboratory at Yale).

It was President Day's predecessor, Timothy Dwight, who recruited Silliman to join the faculty at Yale, at the time comprised of a handful of professors and tutors, none of whom had any experience in the field of science. Dwight's charge, as any college president's is, was to grow the institution, both in terms of new buildings, equipment, and related facilities, but also in terms of its faculty and curriculum in order to attract new students. Dwight was particularly worried about the decentralization tendencies in the early years of the Jefferson administration. Dwight wanted Yale to be more than a regional college: he wanted it to be a magnet strong enough to draw students from far and wide; and that magnet, so he thought, was best found within the new frontiers of science, particularly in the emerging fields of chemistry, geology, and mineralogy.

Enter Benjamin Silliman.

In 1792, President Dwight must have seen something in Benjamin when he matriculated at Yale College at the age of thirteen, becoming a surrogate father to the tall, handsome lad and his older brother, Selleck, Jr., after their father, Gold Selleck Silliman, Sr., who was a close friend of President Dwight, had passed away two years earlier. Known by his classmates as "Sober Ben," Benjamin Silliman was a serious, brooding student, who, in his third year at Yale, wrote the following in his notebook:

> *Journal of a voyage on board the ship assiduity*
> *to the harbor of science, through the ocean of*
> *labour, kept by Benjamin Silliman, commander,*
> *begun July, 1795, and of my college life the 3rd*
> *(aetatis 16 years and 4 mos)*[16]

It's quite a statement from a sixteen-year-old farm lad. It shows several things: that Silliman was serious; that he was eloquent; that he was organized; and that he already had a decided interest in science. The following year, however, Silliman's path was still undecided, and upon graduation, he returned to his hometown of Fairfield, Connecticut, to help his brother fix up the family's Holland Hill farm, whose main

80-acre parcel overlooked Long Island Sound. In search of a livelihood, Silliman accepted a teaching position the following year in a well-known academy associated with an important parish in nearby Wethersfield about the same time that his brother headed to Charleston, South Carolina, to become a private tutor. A year or more later, the brothers reunited in New Haven, where they read law together in the office of Simeon Baldwin, a prominent New Haven lawyer. Several months later, after his brother passed the bar and moved to Newport, Rhode Island, with his wife, Silliman left Baldwin's office to join his friend Charles Denison to read law in the office of Charles Chauncey.

But fate would intervene: in 1800, after one of Yale's longtime tutors accepted the presidency of Middlebury College, Dwight used the opportunity to appoint Silliman to the position of part-time tutor, while encouraging him to continue to study law. After moving into an upper-story room in Connecticut Hall, "Sober Ben" wrote to his mother: "I am as happy as I ever expect to be, perfectly contented and convinced that I am in the road of personal advantage and duty."[17] For the next few years, Silliman divided his time between reading law and hearing student recitations. But President Dwight wasn't through with the young scholar.

Building on an interest in science long held by his predecessor, Ezra Stiles, President Dwight recommended to trustees of the Yale Corporation that they establish a professorship in chemistry and natural history. Part of Dwight's plan to attract students from afar was to keep Yale relevant, and to keep it relevant meant that it had to keep up with its closest competitors. As Harvard, Princeton, and the University of Pennsylvania already had professorships of chemistry, Dwight insisted that Yale follow suit. But this was only one aspect of Dwight's push to make Yale a magnetic intellectual community.

Arguing that the state should not let Yale, "the brightest ornament of Connecticut civilization,"[18] fall into ruin, Dwight convinced the state legislature to appropriate funds to rebuild Old College, the main building on campus, and to construct two new campus buildings: one dedicated to much-needed classroom space and one to use as a student dormitory. With these projects funded, President Dwight returned to his initial call for a new professorship at Yale, which was finally approved by trustees on September 7, 1802. At the same meeting, acting on the recommendation of President Dwight, trustees appointed twenty-three-year-old Benjamin Silliman to the new chair. It was a pragmatic decision: Silliman, a recent alumnus and now part-time tutor, knew the institution (even though he knew very little about chemistry and natural history). Since one of the aims of President Dwight was to create a strong sense of mission among his professors and tutors, it was safer to hire someone familiar with Yale than, as he put it, to hire a foreigner. "A foreigner with his peculiar habits and prejudices," Dwight reminisced later in life, "would not feel and act in unison with us, and that however able he might be in point of science, he would not understand our college system, and might therefore not act in harmony with his colleagues."[19]

Silliman gladly accepted the appointment, though knowing that he had a lot of work to do to prepare him for his first lectures. Over the next year and a half, the young scholar made several trips to Philadelphia, with stops in Princeton to visit John Maclean, Professor of Chemistry. Silliman set out on his travels with the clear expectation that he would see chemistry performed, since up to that point, chemistry was more of an intellectual exercise—something a professor talked about and perhaps demonstrated—than something performed in a laboratory. In Philadelphia, Silliman studied with several eminent scientists at the University of Pennsylvania's medical school, including James Woodhouse, who taught chemistry, and Caspar Wistar, who taught anatomy and surgery.

On April 4, 1804, a year and a half after his appointment by the trustees of Yale Corporation, Silliman gave his first lecture on the history and nature of chemistry. Although nervous about his first course of lectures, Silliman was determined to succeed. His trips to Philadelphia and Princeton paid off in more ways than one: they also gave him the idea for his next effort—establishing a laboratory on the campus of Yale

College in which to perform chemical experiments. Throughout the fall and winter of 1804, Silliman oversaw construction of the laboratory, consigned to the basement of the Lyceum, which stood on the south side of Connecticut Hall. It was a long and tedious project, mainly because the architect had no idea what a chemistry laboratory looked like (the original plan called for a series of groined arches in a subterranean basement that admitted very little light). After Silliman invited President Dwight and members of Yale Corporation to visit the cramped, dimly lit space, which they descended into through an opening in the first floor, trustees approved the changes that Silliman suggested.

But Benjamin Silliman wasn't through. While studying in Philadelphia, he decided that in order to complete his education, he had to visit Europe, in particular England and Scotland, where science education had been steadily gaining ground. After lobbying President Dwight, who then lobbied trustees, members of the Corporation voted to send Silliman to Europe to purchase the requisite materials he desired with a budget of $10,000. On April 4, 1805, almost a year to the date after his first lecture on chemistry at Yale, Silliman boarded the *Ontario* and sailed from New York to Liverpool. Over the next nine months, the inquisitive scholar traveled throughout England and Scotland, meeting with some of the great scientific minds of the day, including John Dalton, James Watt, Henry Cavendish, Thomas Hope, and John Murray.

At this juncture in the Yale enterprise, there was no tension in having a young, energetic professor of chemistry and natural history. Other colleges had similar positions, but they served primarily as islands of knowledge among the principle areas of study that comprised the traditional liberal arts curriculum. In other words, they added to—not detracted from—the core mission of providing a solid, thorough, and prescribed undergraduate collegiate experience. Silliman more than contributed to this mission. Not only were his teaching methods stellar, evidenced by his filled-to-capacity lectures (attended occasionally even by President Dwight), but so too were his many contributions to the Yale enterprise and to the greater academic world in general.

Between his return from Europe in 1806 and the publication in 1828 of the Yale Report, Silliman was involved in a number of important professional activities: he expanded and updated the chemistry laboratory; he brought two important mineral collections or "cabinets" to Yale (one assembled by Dr. Benjamin Perkins of New York; the other by Col. George Gibbs of Newport, Rhode Island); he published the first American study of a meteor (after gaining access to a meteor fragment that fell near the town of Weston, Connecticut); he helped found Yale's School of Medicine, where he taught chemistry and pharmacy over a forty-year period; he founded the *American Journal of Science and Arts* (later shortened to the *American Journal of Science*), and he co-founded the American Geological Society with his longtime friend and supporter Col. George Gibbs.[20]

These milestones only scratch the surface of Silliman's many efforts. By 1828, however, Benjamin Silliman, Professor of Chemistry, Mineralogy, and Geology (a title change effected in 1817) was one of the most well known and widely respected scientists working in the United States. And he was all Yale: he enthusiastically supported the administration of President Dwight and his successors, Jeremiah Day and Theodore Woolsey (and, in turn, was supported by them); he was revered by his students, both undergraduate and advanced; and he was respected—even admired—by his colleagues at Yale. Most importantly, when it came to the Yale Report, Benjamin Silliman supported the Christian-based, traditional liberal arts curriculum that defined Yale's four-year college experience. He saw no conflict between the tradition-bound prescribed curriculum that included the occasional study of science through lecture and demonstration and his interest in teaching applied science to qualified advanced students.

Silliman's pious Christian beliefs were no small thing: the early Yale enterprise demanded it. Silliman demonstrated his devotion to Christian piety early in his career while studying zoology in Philadelphia with Benjamin Smith Barton, the author of the first systematic treatise on botany. When Barton suggested that his students visit a noted natural history collection established by Charles Wilson Peale on a Sunday, Silliman rose in front of his classmates to challenge Barton, declaring: "I regretted to interfere with the wishes or convenience of the class, but that for myself I had other occupations on the day proposed, and if that were to be the time, I must miss the instruction."[21] In other words, the Sabbath is the Sabbath, and should be observed. It was Silliman's unabashed devotion to Christianity that augured well during the early days of Yale's evolution, especially among the overtly religious trustees of Yale Corporation.

AS THE MIDDLE OF the nineteenth century approached, college presidents and faculty were struggling to provide a meaningful curriculum to students in order to satisfy their emerging utilitarian impulses cultivated by the new republic's burgeoning economic growth. Although most colleges chose to do nothing, especially after the 1828 publication of the Yale Report, and others chose merely to tinker around the edges (by offering a small menu of "optional" courses for undergraduates), and only one—Union College under the leadership of Eliphalet Nott[22]—chose a full-blown "parallel curriculum" from which students could choose (one classical; the other practical), most colleges chose to sidestep the question altogether, offering practical courses in science and technology in a separate setting for "advanced" students.

That was the option that Benjamin Silliman and his colleagues at Yale chose, and they did so by creating a program that would not compete with the undergraduate classical program, since that would be a non-starter with the highly ecclesiastical and tradition-bound trustees and faculty of the academic or classical program. In other words, it was a safe choice, one born out of pragmatism, but also one born out of a desire to push the limits of inquiry beyond the confines of the traditional liberal arts curriculum without actually challenging it. It was this impulse that led to the creation of the Sheffield Scientific School, one of the first schools of scientific inquiry associated with an established institution of higher learning. And, for all intents and purposes, it was the brainchild of Benjamin Silliman, Sr. (Yes, he was a father now, and his son, Benjamin Silliman, Jr., would soon join him in the ranks of Yale's faculty.)

Let's start with Charles Warren's article "The Sheffield Scientific School from 1847 to 1947," found in *The Centennial of the Sheffield Scientific School* published by Yale University Press in 1950. Warren begins his overview of the history of the Sheffield Scientific School by reminding us of the context into which the school arose:

> I have been asked to give you a brief outline of the history of the Sheffield Scientific School and of the part it played in the evolution of university education in America. The earlier and firmly established pattern of education in the American colleges was of the inherited, traditional, classical type. Its broad objective was to educate young men for service in "Church and Civil State," and this objective was to be achieved by storing the student's mind with a thorough knowledge of the Greek and Latin languages and of the civilization which they recorded, to provide him with what was regarded as sound and orthodox theology, a correct religious and moral philosophy, and to train his mind in logical thinking. As a part of the process he was exposed to a modest amount of mathematics and natural philosophy, the latter including some physics, astronomy, and natural history. There was no opportunity to study the sciences in any modern sense.[23]

The last line of Warren's introductory statement is curious and can be parsed in several ways. First of all, the phrase "no opportunity to study the sciences" is not completely accurate since from the early 1800s Benjamin Silliman offered courses in chemistry and natural history, specifically geology and mineralogy, as did other small private colleges. Moreover, as early as 1815, Silliman was teaching chemistry and pharmacy in Yale's medical school. It's the latter part of Warren's sentence that is of most interest to us—"in any modern sense"—since it points more toward the way in which science was taught in the early antebellum period rather than to its actual presence in the classical curriculum. Prior to the middle of the nineteenth century most courses of a scientific nature were taught by lecture and demonstration. This was the inherited way of the English collegiate model, influenced by the European Enlightenment and brought into high relief in the lecture circuit of Josiah Holbrook's lyceum movement that was sweeping America during the first half of the nineteenth century.

But there's another overtone of meaning in Warren's concluding statement. From the earliest days of the colonial college, science had always been regarded as the stepchild of the traditional liberal arts curriculum. If it existed in the colonial college curriculum, it did so as an inferior offering to the established program (more "furniture of the mind" than anything else). This is certainly the sense we get when we look at the classical curriculum at Yale in the early decades of the nineteenth century. For a closer look, let's return to Benjamin Silliman, but not to Benjamin Silliman, Sr., but to his first-born son and namesake—Benjamin Silliman, Jr. Born in 1816, Benjamin, Jr. entered Yale as a freshman in 1833 at the age of seventeen. Even before he matriculated at Yale, Benjamin, Jr. helped his father in his professional pursuits, especially in his chemistry lab on the Yale campus. It should be no wonder, then, that upon graduation in 1837, Benjamin, Jr. accepted a position as laboratory assistant and instructor in the department of chemistry, geology, and mineralogy, which his father oversaw. Within a year, he was also helping his father edit the prestigious *American Journal of Science*.

Clearly, Benjamin Silliman, Sr. was the main source of inspiration for the budding scholar, who would slowly become a renowned chemist in his own right (known primarily for his pioneering efforts in the fractional distillation of petroleum). In 1842, a year after he took over his father's editorial responsibilities of the *Journal*, Benjamin, Jr. began teaching chemistry and mineralogy to private students in a spare room in the chemistry lab, doing so at his own expense since trustees of Yale Corporation made no provision for the instruction of advanced students in the sciences. It's not so much *what* Benjamin, Jr. taught, but *how* he taught that brings us back to Warren's statement: he taught not through lecture and demonstration (as his father did in his chemistry and natural history classes), but through hands-on experimentation in the laboratory. This was the "modern sense" to which Warren alluded, becoming ultimately the foundation upon which a new school for scientific study would become established at Yale.

The first stage of this new school began in the spring of 1846, when Benjamin, Jr. wrote a detailed rationale that explained the need for science instruction at Yale for advanced students, suggesting that two new professorships be created: one in applied chemistry; the other in agricultural chemistry and physiology. In full agreement with his son's objectives (even helping him to refine his argument), Benjamin, Sr. presented his son's case to trustees of the Corporation in mid-summer. A decision on the matter was postponed until late summer, when trustees were scheduled to meet in August. At that meeting, two things were resolved. The first was to accept the recommendation that Benjamin Silliman, Sr. presented to trustees of the Corporation: to establish two new professorships. Upon agreeing to do that, trustees turned to filling the positions, relying on Silliman, Sr.'s recommendations: John Pitkin Norton, one of the Silliman's advanced students, was appointed to the professorship in agricultural chemistry and physiology, and Benjamin Silliman, Jr., who had more than proven his worth to the institution, was appointed to the professorship in applied chemistry.[24] This action then posed another question: Where should the professorships, each one for the benefit of advanced students, reside—within the academic faculty of the undergraduate college or within a new department? Without an immediate answer, a committee was formed to study the matter chaired by President Day.

In August of the following year, the committee gave its report, recommending that the Corporation create a fourth department to complement medicine, law, and theology. The unit or division would be called the Department of Philosophy and the Arts. Trustees accepted President Day's recommendation as long as the new department was restricted to "philosophy, literature, history, the moral sciences other than law and theology, the natural sciences excepting medicine, and their application to the arts."[25] But this wasn't the only stipulation. While trustees provided a material site for the new department (the old President's Mansion that was formerly the home of presidents Dwight and Day; which, of course, they rented to Norton and Silliman, Jr. for $150 a year), they declined any financial support for equipment, supplies, and instructor salaries. Funding would be wholly dependent upon student fees and outside donations. It was both a practical and an ideological decision: practically, trustees didn't want the new department to compete for funds earmarked for the academic program; ideologically, the institution took very little interest in science instruction, still seeing it as a stepchild of the established classical curriculum (evidenced further by the fact that instructors in the new department weren't listed in the college catalogue for years to come). Nonetheless, instruction began in the fall of 1847 in the Department of Philosophy and the Arts with the following announcement: "Professors Silliman and Norton have opened a Laboratory on the college grounds in connection with their departments, for the purpose of practical instruction in the applications of science to the arts and agriculture."[26]

Not only did trustees constrain the financial arrangement of the new department, but they also limited the type of students who could attend classes, restricting admission to those who were not members of the undergraduate classes, nor were attending classes in the departments of medicine, law, and theology. (They also made it clear that dismissed students of Yale or any other college were not eligible for enrollment.) In other words, courses were intended for college graduates and "other students of good moral character and adequate preparation."[27] By such constraints, trustees hoped to build a firewall between the traditional liberal arts program (and the professional schools of medicine, law, and theology) and the new department and its science program for advanced students.

From the start, the activities in the chemistry laboratory were a radical departure from the normal affairs of the college. Instead of relying on the age-old lecture-demonstration model, faculty created an experimental testing ground that employed hands-on laboratory methods to encourage students to conduct their own science investigations with minimal interference from faculty. To this extent, students designed and conducted experiments, drawing their own conclusions based on the facts they gathered. Critics, however, thought the entire operation was nothing more than an overrated trade school. But such criticism didn't stop students from enrolling, much to the surprise of everyone.

The success of the science department brought other changes. In 1850, William Norton (no relationship to John Pitkin Norton), an engineer and West Point graduate, and at the time professor of engineering at Brown University in Providence, Rhode Island, was hired as Yale's first professor of civil engineering. Upon accession to the new position, Norton began to develop a systematic program in civil engineering that led, two years later, to a Bachelor of Philosophy degree for students who met the following requirements: they had resided at Yale for two full years, had passed a set of rigorous examinations in three branches of study, and showed proficiency in either French or German.

In 1852, however, the School of Applied Chemistry (as the Department of Philosophy and the Arts was now called) suffered a severe blow when John Pitkin Norton died of pneumonia at the age of thirty just before the fall commencement ceremonies. But the work that Norton had accomplished (with the aid of Benjamin Silliman, Jr., who left Yale several years earlier for a position at the University of Louisville) boded well for the program. In 1854, the School of Applied Chemistry and its counterpart, the Civil Engineering School, were combined to form the Yale Scientific School.

The timing couldn't have been better. Benjamin Silliman, Sr., now in his seventy-third year, resigned two years later after teaching at Yale for more than fifty years. But he did so knowing that his vision of advanced training in the sciences for qualified students had finally been realized. His retirement, however, posed a dilemma for the Corporation: who would succeed him? Upon the recommendation of President Woolsey, who succeeded Jeremiah Day in 1846, and the faculty of the scientific school, trustees split Silliman's position into two professorships: one in geology and mineralogy; the other in chemistry. For the first position, they appointed James Dwight Dana, Silliman's son-in-law and longtime laboratory assistant; for the other position, they appointed Benjamin Silliman, Jr., who had resigned his position in Louisville and returned to New Haven in anticipation of his appointment at Yale.

On August 19, 1856, James Dwight Dana, who had barely settled into his new position, gave the annual commencement address, using it to appeal to alumni and friends of Yale College to support a plan for an expanded program of scientific instruction and research. Under Dana's leadership, faculty of the graduate school embarked on a re-visioning of the school, drafting a document titled "A Plan for a Complete Organization of the Scientific School Connected with Yale College."[28] The plan called for practical training in mining, engineering, agriculture, and manufacturing, as well as graduate work in the natural sciences, and would offer students a Bachelor of Science after two years of study and a Bachelor of Philosophy after three years of study.[29] The next step was fund-raising, which the faculty did, but not through Yale's alumni or even the state of Connecticut; they did so by soliciting private donations, assigning those who contributed $5,000 or more to a board of visitors with the power to make recommendations to the scientific school's faculty and to the college president.

In their first year of fundraising, the school raised over $25,000. Of that sum, $10,000 came from New Haven railroad investor and financier Joseph E. Sheffield, who was the father-in-law of John Addison Porter, John Pitkin Norton's successor as professor of agricultural chemistry and physiology. Deeply interested in promoting scientific and technical education, Sheffield would continue to underwrite the work of the Yale Scientific School, contributing not only money ($50,000 in 1859 alone to endow three professorships in chemistry, metallurgy, and engineering), but also material resources (a large building at the head of College Street that was remodeled and outfitted with scientific equipment and named Sheffield Hall in his honor). By 1861, trustees of Yale Corporation were so impressed by Sheffield's munificence, by then totaling over $1,000,000, they decided to rename the institution the Sheffield Scientific School.

In retrospect, how could Yale College, the institution that issued the Yale Report of 1828 in defense of the traditional academic liberal arts program, host one of the leading scientific schools in the nation? The answer resides in the fact that trustees of Yale Corporation and its academic faculty, including Benjamin Silliman, Sr., kept the two entities separate. Not only did the Corporation decline to offer financial resources to the scientific school, but as we've already seen, it also restricted admission to the school to any well-qualified candidate who was *not* part of Yale's traditional undergraduate program or its professional schools. In fact, it was not until 1888 that undergraduates could attend classes in the Sheffield Scientific School at all. In other words, by establishing a separate scientific school, trustees of Yale Corporation hoped to achieve three distinct goals: maintain the traditional undergraduate curriculum; satisfy the demands of those who advocated for more instruction in the sciences; and, most important of all—at least for trustees of Yale Corporation—continue to receive donations from wealthy benefactors. The latter is illustrated in the following case: when a wealthy donor offered the institution $20,000 to endow a chair of botany in 1864, trustees graciously accepted the donation, but made it clear to the donor from the outset that such instruction had no place in the classical liberal arts program, but was more than appropriate for the scientific school. In other words, by keeping the programs separate, Yale could have its cake and eat it too.

Eight years after his retirement in 1856 as Yale's first professor of chemistry, geology, and mineralogy, Benjamin Silliman, Sr. died. He was eighty-five years old. Writing for the centennial celebration of the Sheffield Scientific School—"Old Sheff," as it was affectionately called—Leonard Wilson offers this tribute to the man who not only made science an essential part of general education, but also planted the seed of graduate education that would soon extend from the sciences into every other field of learning:

> Benjamin Silliman was a large-minded and tolerant man. He encouraged, promoted, and assisted the efforts of John Pitkin Norton and Benjamin Silliman, Jr. to establish a scientific school at Yale in 1846. The gradual academic evolution of that school into the Yale Scientific School in 1854 and the Sheffield Scientific School in 1861 occurred with Silliman's unfailing encouragement, advice, and support. The beginning of graduate scientific education at the Sheffield Scientific School in 1861 was a culmination of Silliman's lifelong effort to establish both science teaching and scientific research in the United States.[30]

In 1846, the Sheffield Scientific School was little more than an idea in the minds of Benjamin Silliman and his son (and the few dedicated students who did advanced work in chemistry under their direction). In that same year, in another part of New England, Edward Charles Pickering was born. Eighteen years later, the same year in which Benjamin Silliman, Sr. died, Edward Pickering was in his third and final year at Harvard's Lawrence Scientific School. He would graduate *summa cum laude* at the end of the year with a degree in physics, but only first having tried his hand at chemistry.

If Edward Pickering had grown up in New Haven and attended Yale, he still would have been in his junior year when Benjamin Silliman, Sr. died, only he would have been enrolled in the undergraduate liberal arts program, since Yale offered applied science instruction only to well-qualified "advanced" students. Ultimately, the scientific school would expand its programs and student base, but for now—in the middle of the nineteenth century—it was moving slowly, keeping undergraduate and graduate education distinct. Harvard, on the other hand, would choose a different path.

CHAPTER 6 | *An Impenetrable Thicket*

> *We have now learnt that as many years are passed in our schools, and colleges, and professional preparation, as are passed in the same way, and for the same purpose, in the best schools in Europe, while it is perfectly apparent that nothing like the same results are obtained; so that we have only to choose whether the reproach shall rest on the talents of our young men, or on the instruction and discipline of our institutions for teaching them.*[1] —George Ticknor

In 1630, the Massachusetts Bay Colony founded a small town eight miles north of Boston and named it Newtowne. It was intended as an enclave for Boston's wealthy families. Removed from the noise and grime of Boston, a rough-and-tumble port city, those with significant means sought a more rustic, pastoral life. At the direction of the Great and General Court of the Massachusetts Bay Colony, city fathers changed the name of the small town to Cambridge in 1637. It was both a tip of the hat to England's Cambridge University where many of the early colonists had studied and an acknowledgment of the hopes and aspirations of the colony's newly established college.

Founded in 1636, New College was the first corporation chartered by the Massachusetts Bay Colony. When its doors opened two years later New College had a faculty of three—a schoolmaster and two tutors—that offered a classic Latin grammar education based on the English university model. From the start, Puritan fathers expected New College to provide future church and civic leaders an education comparable to its English counterpart. But New College got off to a rocky start when newly appointed schoolmaster Nathaniel Eaton was fired during his second year in office for beating his students and one of his tutors. Classes resumed in the summer of 1640 under the direction of Henry Dunster, Eaton's replacement and officially New College's first president, only the college was no longer called New College. In 1639, the year prior to Dunster's appointment, the General Court renamed the school Harvard College.

It was another tip of the hat, this time to the college's benefactor, John Harvard, a resident of nearby Charlestown, who bequeathed his library and half of his estate to the fledgling institution. Harvard, who held a master's degree from Emmanuel College in England, immigrated to the colonies in 1637 with his wife.[2] They settled northeast of Boston in Charlestown, where Harvard became a teaching elder and assistant preacher at First Church. He had been at this work for little more than a year when he contracted tuberculosis and died on September 14, 1638. Before he died, the thirty-year-old minister, who had inherited a small fortune from his family, dictated his last will and testament to his wife, bequeathing his library of four hundred books and half of his property to New College.

So moved were members of the General Court by John Harvard's bequest that on March 13, 1639, they decreed that New College be renamed Harvard College, reiterating their hope that—in the English sense of the word—the institution become a real college: "a society of scholars, where teachers and students lived in the same building under common discipline, associating not only in lecture rooms but at meals, in chambers, at prayers, and in recreation."[3] Their hope came to fruition in 1642 when the first college building was built, uniting under a single roof teachers and scholars. It was an essential arrangement, borrowed from English collegiate life, where teachers and scholars lived together sharing meals, chambers, prayers, and recreation. It was, in effect, total immersion in a setting devoted to learning and to the cultivation of moral character.

Old College was a four-story wooden structure built in the shape of an "E" with two open quadrangles. The first floor contained a large assembly room where teachers and scholars met for prayers, meals, and college exercises. Adjoining the large assembly room were several utility rooms for storing, preparing, and serving food. The library took up the second floor along with alcoves for quiet study. The remaining floors were reserved for the living quarters of masters, tutors, and scholars. Unfortunately, the building was poorly designed and built, requiring a constant stream of repairs. Three years after the General Court renamed New College, Harvard graduated its first students—nine in all.

Although founded primarily as a seminary for the training of clergy, Harvard College would wrestle with the place of religion in both its governance and its curriculum. The tension was the result of an ongoing internal struggle between the orthodox and liberal wings of Puritanism, between Congregationalists who wanted Harvard to be a seminary for the education of ministers and Unitarians who wanted Harvard to be a center of liberal arts education. The first break between the orthodox and liberal wings of Puritanism occurred in the late 1600s, when, in an attempt to reorganize the New England colonies, the English monarchy revoked the charters of both the Massachusetts Bay Company and Harvard College. To help negotiate a resolution to the crisis, Harvard's president, the influential orthodox Puritan leader Increase Mather, was called to London, leaving the administration of Harvard to two of its most capable tutors, John Leverett and William Brattle, both of whom were Unitarians and supported by Boston's influential merchant class.

While Mather was abroad, the Unitarians orchestrated a coup, forcing Mather to resign his presidency upon his return to America and installing the liberal Samuel Willard in his place, who was succeeded shortly afterwards by Leverett upon Willard's death. Leverett's appointment came with the backing of both Boston's merchant class and the colonial governor. The resolution of the president's position was a solid win for open-minded, liberal Unitarians. More than anyone else, the "Great Leverett," as he would be remembered, not only shielded the college from the constraining hold of Puritan orthodoxy, but also firmly established Harvard's liberal arts tradition.[4]

In a strange way, however, one of the most influential forces that shaped Harvard's curriculum in its early days was not the appointment of John Leverett, or any of his liberal-minded successors: rather, it was a cataclysmic event—the Great Harvard Hall Fire of 1674. Built at the edge of Harvard Yard after Old College collapsed in the early 1670s, Harvard Hall contained a number of meeting rooms, the college library, and a growing collection of scientific instruments. With a smallpox epidemic raging in Boston, members of the Great and General Court removed themselves to Harvard, using the upstairs library in Harvard Hall as their meeting room.

On a bitter night in late January, with temperatures plummeting and snow swirling outside, members of the General Court huddled inside Harvard Hall, warmed by the fire blazing in the hearth on the building's second floor. After meeting for the better part of the evening, the General Court adjourned and members retired for the night. Some time before midnight, several townspeople noticed smoke coming from Harvard Hall. Within minutes, the building was an inferno. Although Harvard Hall could not be saved, members of the General Court and residents of Cambridge pitched in to help keep the fire from spreading to nearby buildings. Fortunately, Massachusetts Hall, Stoughton Hall, and newly constructed Hollis Hall were spared Harvard Hall's fate—total destruction.

Later, it was discovered the fire that members of the General Court had started in the library's hearth on the second floor had not been completely extinguished upon the Court's adjournment. Apparently, the remaining embers set the floor beams on fire, which in turn set the entire building on fire. Among the charred embers of the building were all of John Harvard's books, all but one that is. Spared from the conflagration—only because it was checked out at the time (and overdue at that)—was the fourth edition of John Downame's *The Christian Warfare Against the Devil World and Flesh* published in 1634. This, and four hundred other books from the library that were either out on loan or newly acquired and not yet unpacked (and, thus, quickly hustled out of the building before they were consumed), was all that was spared. In the aftermath of the fire, the General Court took full responsibility, agreeing to pay for the loss of the building and its contents and all of the possessions that students and staff lost. The General Court even agreed to donate a water engine to Harvard College to be used for future firefighting efforts. The new building was built on the same location in Harvard Yard as the one that had succumbed to the fire. Completed in 1766, the building opened with a number of new fire regulations put in place by Harvard's Board of Overseers.[5]

The real impact of the Great Harvard Hall Fire of 1674, however, wasn't the new set of fire-prevention regulations; it was the outpouring of support, both in terms of financial donations and in-kind book donations, that Harvard College received from nearly three hundred alumni and friends of the college, among them John Hancock, Benjamin Franklin, William Dummer, former Lieutenant Governor of the Massachusetts Bay Colony, and Thomas Hollis, a wealthy Englishman who donated most of his resources to promoting liberal education in the colonies.

By the time the new building was completed, the library collection far surpassed the one before the fire. Located in the upper west chamber of the building, books were arranged on shelves within alcoves named after their benefactor. But it wasn't the amount of volumes that impacted Harvard's curriculum and the experience of its students; rather, it was the broader selection of books that allowed—even encouraged—faculty and students to read in an ever-widening secular fashion. This was the real impact of the Great Harvard Hall Fire of 1674. A relative calm would settle over Harvard for the next hundred years. It was a time of peace and prosperity as the small colonial college found its footing in the new world, until the Sons of Liberty rose up against their British oppressors.

ALTHOUGH THE OUTBREAK OF the American War of Independence was a setback to the growth and prosperity of Harvard, it was a temporary setback at best, as it—or rather, its outcome—spurred Harvard to new heights as the spirit of "republicanism" swept the country. Learned citizens understood that the survival of the new republic depended upon the virtue of its citizens; that is, the capacity of individuals to put public good above their own personal interests. This was the generous and liberal ideology of the European Enlightenment. For education, the inculcation of virtue meant increased attention to history, political theory, and law, as well as the study of modern languages, especially French, which at the time was not only the language of the courts of Europe, but also, along with German, the academic language of leading scientists. It also meant increased attention to the professions, and not just to medicine, law, and theology, but also to architecture, engineering, and other trade-oriented professions.

Although the fight for independence interrupted the growth and development of Harvard during the Revolutionary War period, it did not interrupt the struggle between religious conservatism and liberalism. Less than thirty years after gaining independence from Britain, citizens of the new republic experienced a revival of conservative religious views that swept the country in the first decade of the nineteenth century. For many, it was the long-awaited triumph of religious conservatism over republican liberalism. Called the Second Awakening, religious fervor reverberated throughout the country, except in Boston, a city whose churches were firmly in the hands of liberal Unitarian ministers.[6]

Harvard, too, in neighboring Cambridge, was buffered from the revivalist sentiment sweeping the country, as its administration and principal backers were comprised mostly of Boston's liberal-minded merchant class. This was demonstrated after the sudden death of President Joseph Willard in 1804. In a fight reminiscent of the earlier struggle between the conservative and liberal wings of Puritanism that resulted in the ouster of Increase Mather and his followers, Harvard's governing board ignored the pleas of conservative Congregationalists and appointed Samuel Webber, a liberal Unitarian minister, over and above acting president Eliphalet Pearson, a staunch Congregationalist who held the position after Willard's death. It was another notch in the belt for liberal ideology. Four years later, the evolution of Harvard as a predominantly liberal Unitarian college was accelerated when John Thornton Kirkland, a distinguished Unitarian minister, was elected to succeed Webber. Kirkland's presidency signaled the beginning of Harvard's Augustan Age, an age of curriculum expansion and innovation, all generously funded by Boston's wealthy and liberal-leaning merchant class.

As many a Harvard president had, Kirkland descended from admirable stock: in his case, from patriot Miles Standish. Although bearing an amicable, even sweet personality, Kirkland was serious in attitude and bookish from a young age, preferring to read a book rather than to go out and play with his classmates. In 1784, at the age of thirteen, Kirkland's father brought him to Andover, Massachusetts, and enrolled him in Phillips Academy, which, at the time, was under the direction of Eliphalet Pearson, the same Eliphalet Pearson who, as Harvard's acting president, was displaced by Samuel Webber. Kirkland excelled at Andover, having been taken into the house of Samuel Phillips, the school's founder. Two years later, Kirkland moved to Cambridge, where he attended Harvard, graduating with distinguished honors in 1789 at the age of nineteen.

After serving as a Harvard tutor for several years, Kirkland became the pastor of New South Church, one of Boston's most theologically liberal but politically conservative Unitarian churches. As he did at Andover and at Harvard, Kirkland excelled in his new position, quickly becoming a respected and influential member of Boston's professional class. Although broad-minded and universal in his knowledge, Kirkland could also be quite conservative, especially in his view of education. A gentleman and a scholar, Kirkland believed that the foundation of education was first and foremost a thorough understanding of ethics, with philosophy next, followed by the natural sciences. This was not unusual for Unitarians; as descendants of strong Puritan stock, they had a strong sense of moral direction and an interest in maintaining the economic and social status quo.

The combination of Kirkland as a devout Christian and scholar made him a perfect choice for the presidency of Harvard College after Webber's death in 1810. Trustees of Harvard Corporation were not unfamiliar with Kirkland; in fact, they had considered him to succeed Joseph Willard six years earlier, but fearful of aggravating the already strained relations with the college's conservative Congregationalists, they chose Samuel Webber instead. When Webber died six years later, the Corporation overcame their hesitancy and elected Kirkland its next president.

With the backing of the Corporation, Harvard's Board of Overseers, and Boston's merchant class, Kirkland launched a full agenda of reforms. He oversaw the construction of several buildings; cleaned up Harvard Yard, which still had echoes of its cow-yard watch days; acquired permanent quarters for the medical school in Boston; and opened a school of law and a school of divinity in Cambridge. Additionally, through numerous private donations, Kirkland enlarged Harvard's endowment, established a student scholarship fund, increased faculty salaries, and established over a dozen new professorships, and all within his first seven years of office. The result was palpable. By 1818, Harvard's enrollments had doubled, and for the first time since the 1770s exceeded that of its closest competitor—Yale College in New Haven, Connecticut. But even during what has been called Harvard's robust Augustan Age, Kirkland demonstrated a cautious, even conservative approach to curriculum reform by placing most of his emphasis on strengthening the existing professional schools and leaving other much needed reforms, like the development of practical or applied science, for another day.

It wasn't that he was unaware of reform efforts at other colleges; he was just stubbornly resistant. When George Ticknor, Harvard's Smith Professor of French and Spanish Languages and Literatures, proposed a number of curriculum reforms in 1821, such as introducing a limited elective system, grouping students by proficiency rather than alphabetically by class, teaching by subjects rather than by books, and replacing daily recitations with lectures and written exams, Kirkland chose to side with dissenting faculty rather than with Ticknor. Kirkland feared that Ticknor's aim was not merely to add variety to a heavily classical orientation, but to break up completely the lock-step curriculum adopted by the institution's founders. Kirkland, though a liberal Unitarian, was at heart a conservative educator wedded to the English scholastic model.

The president's undoing, however, came in an unexpected manner. A kindly, fatherly figure, dearly beloved by the student body, Kirkland showed little backbone when it came to disciplining unruly students, and Harvard, like other colleges, had their fair share. The post-Revolutionary War period was a heady time, with the spirit of republicanism thick in the air. It was certainly thick in the hearts and minds of students who were coming of age in a country that itself was still coming of age. Restlessness, even rebellion, marked the times at most private colleges. Kirkland's refusal, or perhaps his inability, to deal with student rebelliousness raised red flags among members of Harvard Corporation and its Board of Overseers. One person in particular, Nathaniel Bowditch, a long-time member of the Board of Overseers and newly appointed member of the Corporation, was so alarmed by the besmirching of Harvard's reputation by unruly student behavior—from all-night carousing to physical attacks on faculty—that he asked for and got a full review of Kirkland's presidency. What he found in the process brought Kirkland's administration to its knees: despite increased enrollments, Harvard was not only in dire financial straits due to fiscal mismanagement, but its reputation was suffering as well, so much so that local ministers advised young men to go elsewhere, somewhere less costly, less scandalous, and doctrinally more sound, like Bowdoin, Williams, or even, heaven forbid, Yale. Bowditch and Kirkland finally came to blows at a Corporation meeting on March 27, 1828. Seeing the writing on the wall, Kirkland threw in the towel and resigned the following day.

The upshot was that Ticknor's reforms withered on the shelf while faculty support for the classical curriculum became emboldened. Besides, the new president would have to deal with student behavior first, not curriculum reform. The new president needed other qualities as well, in particular, an understanding of organizational structure and finances. That was why trustees of Harvard Corporation turned to Josiah Quincy, the former mayor of Boston, for its next leader, rather than one of their own. Quincy, a stocky, energetic man in his late fifties, whose background was in banking, insurance, and real estate, would also be the first Harvard president without a ministerial connection to Boston's elite churches. But that didn't mean that he didn't have a connection to Harvard. He did, as alumnus, benefactor, and member of the Board of Overseers. He was a natural choice and on January 15, 1829, less than two weeks after Quincy relinquished his elected mayoral office, Harvard Corporation unanimously elected Quincy the seventeenth president of Harvard College.[7]

JOSIAH QUINCY THRIVED IN the public eye. For most of his adult life, the square-jawed, stocky Quincy was an uncompromising public servant, acknowledging as much in his acceptance speech when he stated: "I recognize the right of society to command my services; and I accept the appointment, as a duty, which I have no authority to decline."[8] His first act as president was a five-week tour of several Eastern colleges, including Yale, Columbia, and the University of Pennsylvania. It was an inspired, even necessary act, for Quincy had no prior academic administrative experience, nor a deep understanding of collegiate life. What he found as a result of the tour was a confirmation of his suspicions that curriculum innovation without attention to student discipline, as was the case during Kirkland's tenure, was doomed to failure. Any effective educational program had to be based on a sound system of student discipline.

Upon returning to Harvard, Quincy took on an authoritative parental attitude toward student unrest. After all, this was how he brought municipal departments under his control as mayor of Boston. With a reputation for being a no-nonsense enforcer of civil law, both as a municipal judge and as the city's mayor, Quincy struck fear into the hearts and minds of Harvard's student body, laying down the law with students at the very outset of his term in office, going so far as to warn them that riotous behavior would not be tolerated, that it would be dealt with harshly, and that, if need be, perpetrators would be turned over to the municipal police and courts. Quincy's hardline stance worked for a while until the inevitable tension between the president and the student body finally erupted into open warfare in 1834 during Quincy's sixth year in office.

The precipitating event was the verbal attack of a Latin tutor before his students (and, later, the setting on fire of the tutor's recitation room by a group of students who sided with their classmate). In response to Quincy's swift punitive action against those involved, additional student unrest followed, first by the lower classes, then by the entire student body. As he did while mayor of Boston during difficult times, Quincy kept to his hardline policies, assuring members of the Corporation and the Board of Overseers, that his was the right course of action: "I am not a man to be frightened from a post of duty and usefulness, the harder the tempest rages, the tighter I shall stick to the rudder."[9] Not everyone agreed with Quincy's harsh tactics. Faculty members were put off by Quincy's top-down, authoritarian attitude, as were several members of the Board of Overseers. It was only after the esteemed John Quincy Adams made it clear that students—not the president—were responsible for the Rebellion of 1834, as the episode was called by the local press, that tempers quelled and students backed down.[10]

The resolution of the student riots of 1834 had several long-lasting results—some intentional, others not. First of all, due to Quincy's heavy-handed tactics, it became quite clear that control of Harvard College rested not with its students, its faculty, or even its Board of Overseers, but with its president and members of Harvard Corporation. In the short run, however, the student rebellion delayed any hope for curriculum reform. Maintaining student discipline in a unified, lockstep curriculum became Quincy's primary goal, at least for the near future. But in the long run, ironically, the rebellion—or at least the resulting clampdown on student dissent—made reforms possible. Once Quincy had gained control of the student body, he could begin to turn his attention to the curriculum.

To the dismay of reform-minded members of the Harvard faculty, Quincy turned out to be a traditionalist who believed in the stultified English collegiate model. When a loosening of the lock-step curriculum was suggested by several faculty, Quincy balked, reiterating what he thought was necessary for Harvard to become a beacon in the world of higher education: stiffer entrance requirements, a more demanding classical curriculum, and more exacting degree standards. But if his view of the college were pursued, it would be at the peril of its own survival. In the aftermath of the War of Independence and on the cusp of the Industrial Revolution, America was changing—and so were its colleges. However, short of making Harvard an open-enrollment trade school, which some alumni and faculty were advocating, Quincy relented and began to listen to the voices of reform.

The two loudest voices came from German-born and educated Charles Beck, Professor of Latin Language and Literature, and Benjamin Peirce, Perkins Professor of Astronomy and Mathematics. Beck had already begun asking for changes to the curriculum several years prior to the 1834 student rebellion. Shortly after he joined the Harvard faculty in 1831, Beck proposed the creation of a two-year philological seminary that would train students in the nuances of classical scholarship, with a focus on the nature, structure, and historical development of Latin. It was a unique form of professional training for students who wished to teach classical literature, unique in that students began the program during their senior year and completed it at the end of an additional fifth year. It was an early prototype of postgraduate study in a field other than medicine, law, or theology. Beck's idea was also unique in that it called for a program not only distinct from the regular college curriculum, but also staffed with its own faculty.

Although hesitant to upset the lock-step classical curriculum for which Harvard was founded and known, Quincy ultimately supported the idea, as it would provide a highly qualified pool of classical scholars that the college could tap for future tutors and professors. With the president's support and the Corporation's approval, the first group of six students began their seminary experience in the spring of 1832. But the program ran aground almost immediately. Although it was a two-year program, many students chose not to return at the end of the first year since, technically, it was the end of their four-year undergraduate program and, as such, were eligible to receive their baccalaureate degree (which many of them took and then left to teach in a secondary school, avoiding not only an additional year of study, but also an additional year of college expenses). The final nail in the coffin came after Beck asked for, but was denied, scholarship money for the remaining students enrolled in the program. The program closed shortly after that.

A year after the student rebellion of 1834, Benjamin Peirce joined the Harvard faculty as Professor of Mathematics and Natural Philosophy (later named Perkins Professor of Astronomy and Mathematics). Upon his appointment, Peirce began a fierce attack on the prescribed classical curriculum, proposing the radical notion that all mathematics courses beyond the freshman year be optional. Peirce believed that the required three years of mathematics at Harvard was a complete waste of time for most students, calling it "a veritable forced march ending in what was for most the impenetrable thicket of differential calculus."[11] Peirce believed that student interest and ability should drive the selection of classes and not the imposition of a fixed, prescribed, one-size-fits-all curriculum.

Whereas Quincy supported Beck's seminary proposal because it extended rather than undermined the classical curriculum, he tabled Peirce's suggestion for its more radical and disruptive effect. Quincy was unwilling to allow the student body to be divided so radically by student proficiency. After all, why should certain areas of the curriculum be proscribed to all but certain carefully selected students? This was too radical a notion for the conservative leaning and classically trained Quincy. He was not yet prepared to adopt the model of individual scholarship and performance over the tradition of uniform instruction at Harvard. But Peirce persisted and in 1838 offered another proposal, one that combined his original elective program with Beck's professional training model. Peirce proposed three tiers of course offerings: a one-year course in practical mathematics of use to surveyors, navigators, and others working in the technical trades; a more theoretical one-year course designed to train future mathematics instructors, especially at the secondary school level; and a three-year program for the most capable who wanted to obtain a professorship in mathematics.

Recognizing the potential behind Peirce's new proposal, Quincy supported it: the changes would alter, but not displace the classical curriculum, and, more importantly, they would not require additional institutional funds to implement them. At the end of the 1838 academic year, Quincy persuaded members of the Corporation to adopt Peirce's plan on a two-year trial basis. Several months after the Corporation's approval, the heads of both the Latin and Greek departments followed suit and implemented their own modified elective system. Now there was no turning back. Since instruction in Latin, Greek, and mathematics represented three of the fundaments of the classical curriculum, the elective system was here to stay. At least that was what it seemed by 1843, when all courses beyond the freshman year at Harvard were optional. And they would remain so, but only as long as Josiah Quincy remained Harvard's president.[12]

As insurgent as they appeared at the time, the structural changes to the classical curriculum proposed by Beck and Peirce did not address the larger issue of scientific and technological training at Harvard. As important as these areas were, they would need more than the support of several faculty members; they would need the president and members of the Corporation to lead the way—as well as the deep pockets of an outside donor or two.

IN THE SPRING OF 1846, Edward Everett succeeded Josiah Quincy as Harvard's eighteenth president. It was the same year that Benjamin Silliman and his son began dreaming of a scientific school at Yale. Everett's appointment boded well for the future of science education at Harvard, even though in his inaugural address he stood firmly behind the value of a traditional liberal arts education ferried across the Atlantic by his conservative English forebears.

The Hon. Edward Everett, LL.D. came to the presidency of Harvard College with more than enough credentials to support his candidacy. Born in 1794 in Dorchester, Massachusetts, Everett was the son of a minister and as such received the educational training commensurate with someone of his lineage. He attended and excelled at several preparatory schools, including Exeter Academy in New Hampshire, prior to matriculating at Harvard College in 1806 at the age of thirteen. Four years later, Everett graduated with high honors and matriculated immediately at Harvard's divinity school. Upon graduation in 1813, Everett ascended to the pulpit of Brattle Square Church in Boston, which was a stronghold of Unitarianism for most of the nineteenth century. Two years later, Everett assumed the newly established Eliot Chair of Greek Literature at Harvard, funded generously by Boston banking executive Samuel Eliot under the presidency of John Thornton Kirkland.

After traveling to Europe for additional studies (with friend and fellow Bostonian George Ticknor, who had just accepted the Smith Professor of French and Spanish Languages and Literatures at Harvard), Everett took up his teaching duties in the fall of 1819, remaining at Harvard until 1824, after being elected to Congress, where he represented Middlesex district in the U.S. House of Representatives. A staunch supporter of President John Quincy Adams, Everett remained in Congress until 1836, leaving to serve as Governor of the Commonwealth of Massachusetts, a position he held for only one term. Although a one-term governor, Everett oversaw the creation of the State Board of Education (affirming the appointment of reform-minded Horace Mann as the nation's first State Secretary of Education) and the establishment of a system of Normal Schools to educate primary and secondary teachers.

After losing the governorship by the slimmest of margins in the 1840 election, Everett sailed to Europe with his family, only to be enlisted by the Harrison administration to serve as U.S. Ambassador to Great Britain. Upon his return in 1845, trustees of Harvard Corporation and members of the Board of Overseers recruited Everett to succeed Josiah Quincy. But of all his positions, the least satisfying and effective was his presidency of Harvard, which lasted only three years. In a tribute to Everett upon his death in 1865, Rev. John Todd offered this explanation of Everett's disappointing tenure as president of Harvard College: "His mind, high-toned by nature, refined by culture, long accustomed to the courtliness of the finest society in the world, and the dignity of diplomatic circles, could not adapt itself to the management of young men in that half-fledged state,—when they have ceased to be boys, and have not yet learned to be gentlemen."[13]

As Paul Varg explains in his biography *The Intellectual in the Turmoil of Politics*, it wasn't only student misbehavior that undid Everett's presidency; it was Everett's attempt to transform his alma matter into a university similar in organization to what he had experienced at the University of Göttingen some thirty years earlier. According to Varg, Everett wanted Harvard to become "a center of learning offering a broad program including science, a true university enriched with professional schools staffed by men dedicated to research, and it was to become the gateway through which young men acquired a reverence for learning and a sense of duty rooted in Christian principles."[14]

Everett addressed the idea in his inaugural address delivered on April 30, 1846, to an illustrious body of students, faculty, trustees, overseers, and distinguished alumni, but only after he reminded attendees of the history of "The University at Cambridge," as Everett called Harvard, citing a reference from the constitution of the Commonwealth of Massachusetts. From there, Everett made several observations concerning the difference among American, English, and German institutions of higher learning, heaping praise on the latter for their embrace of the university concept and professional schools beyond the usual triune of medicine, law, and theology, ending his meditation with a question:

> [Has not the time arrived] when a considerable expansion may be given to our system, of a twofold character; *first*, by establishing a philosophical faculty, in which the various branches of science and literature should be cultivated, beyond the limits of an academical course, with a view to a complete liberal education, and, *secondly*, by organizing a school of theoretical and practical science, for the purpose especially of teaching its application to the arts of life, and of furnishing a supply of skilful [sic] engineers, and of persons well qualified to explore and bring to light the inexhaustible natural resources of the country, and to guide its vast industrial energies in their rapid development.[15]

After raising the question and posing two possible solutions, Everett immediately put the question aside and launched into a treatise on the value of the general system of classical training, of which he saw three broad aims:

> *First*, the acquisition of knowledge in the various branches of science and literature, as a general preparation for the learned professions and the other liberal pursuits of life;— *Secondly*, in the process of acquiring this knowledge, the exercise and development of the intellectual faculties, as a still more important part of the great business of preparation;—and, *Thirdly*, the formation of a pure and manly character, exhibiting that union of moral and intellectual qualities which most commands confidence, respect, and love.[16]

With these three points, Everett could have been speaking in support of the Yale Report that so eloquently and uncompromisingly defended the classical liberal arts curriculum. Indeed, most of Everett's speech addressed these points, starting with a justification of the subjects that made up the classical curriculum, arguing against those who said that much of what is studied is a waste of time as such studies provided little useful application to contemporary life. Everett, the master orator, turned the question around: "It is not that the studies pursued at the university are of no use to life, but that we make no use of them."[17]

Having addressed the first of the three aims of higher education, Everett then turned to the need to discipline the mind—to imbue it with the cardinal powers of attention, perception, memory, judgment, abstraction, and imagination. For this, Everett cited the study of language, modern and ancient: "That there is something in the study of language extremely congenial to the mental powers of most men is sufficiently shown in the almost miraculous facility with which, even in infancy, the vast circle of a language is substantially mastered."[18] He did not leave out the exact sciences as contributing to disciplining the powers of the mind, but limited his examples to the two most traditional subjects of the classical curriculum—mathematics and astronomy.

Finally, Everett arrived at the third aim of higher education—developing moral character—calling it "the most precious endowment of our fallen nature."[19] Reflecting the strong Puritan roots of Harvard's early history, and his own religious upbringing, Everett quickly dug into the heart of the matter: "I know of no reliable foundation but sincere and fervent religious faith, founded on conviction, enlightened by reason, and nourished by the devout observance of those means of spiritual improvement which Christianity provides."[20] Everett's emphasis on the development of moral character through Christian piety certainly was not out of keeping with Harvard's past (despite the many immoral acts students perpetrated upon each other, their tutors, and even the president himself). But let us not confuse Christian piety with religious training: although a religious man himself, Everett wanted to decouple Harvard from its divinity school, which, according to Everett, was the weakest professional school at the institution. However, like many of his initiatives, his effort to separate the divinity school from Harvard—both administratively and geographically—met with strong resistance and was ultimately vetoed by members of the Corporation.

Although Everett did not have the full support of the Harvard community behind him, he did have a small contingent of professors who supported his ideas and with whom Everett felt intellectually at home. At the core of this group was Benjamin Peirce, who, along with his proposal to introduce an elective option into the mathematics program, had been urging members of the Corporation for several years to establish an extra-faculty school of science, a theme picked up by Everett in his inaugural address when he questioned whether or not it was time to organize a school of theoretical and practical science.

The idea for a school of science was not Peirce's idea alone. While a student at Harvard, Peirce came under the influence of Jacob Bigelow, who held the first Rumford professorship of applied science.[21] Bigelow assumed the position in 1815, one year after it was endowed by Sir Benjamin Thompson, a.k.a. Count Rumford, an American-born military officer and British loyalist who moved to England (where he was knighted for his contributions to science), and later to Bavaria (where he was given the title Count of the Holy Roman Empire for his service to the Duke of Bavaria). Although Thompson spent much of his adult life outside of his native country, Thompson never forgot his roots,[22] bequeathing a large portion of his estate to Harvard for the purpose of establishing a professorship in the application of science to everyday practical problems, following his lifelong pursuit of scientific discovery and invention (for which he was admitted to the Royal Society of Great Britain and the American Academy of Arts and Sciences). But it is Bigelow, not Benjamin Thompson, with whom we are concerned, for it is Bigelow who served as Rumford Professor of Applied Science from 1815 to 1827.

A successful physician and part-time chemist, Bigelow showed an insatiable curiosity for how things worked, as well as an interest in the link between practical knowledge of scientific principles and the improvement of the human condition. In this sense, as Mary Ann James notes in her treatise on science education at Harvard during the early nineteenth century, Bigelow's views were not just academic; they were also highly political. To Bigelow, elitist philosophies of the past that were restricted to the wealthy few and offered "no practical benefit or means of serving others when completed" seemed not only out of place in an emerging democracy, but also contrary to it.[23] This sentiment is found in the message that Bigelow delivered to the senior class in 1824, when he argued, "technical knowledge was just as important a study as any branch of literature, and as valuable to them as the study of any language."[24]

Cut from the same cloth as other mid-nineteenth-century liberal-minded reformers dedicated to bringing institutions of higher learning in line with the growing industrial and technological needs of society, Bigelow argued for the modernization of Harvard's curriculum, averring that what America needed was modern knowledge, modern languages, and modern skills that best fit with the needs of its people. Instead, as James points out, "young men were presented with an educational program transferring the elitist symbols and habits of a privileged English class to a new American generation."[25] It was not what a fledgling democracy needed or expected from its educational system. This sentiment is precisely what motivated Thomas Jefferson's ideas for educational reform in the Commonwealth of Virginia, ideas unfortunately that never fully came to fruition in the planning and execution of the University of Virginia.

Bigelow resigned from his duties as the first Rumford Professor of Applied Science in 1827 (though he maintained an affiliation with Harvard's medical school for the remainder of his life). His repeated calls for the elevation of science as a respected academic discipline within the walls of Harvard receded into the shadows, but not for long: four years after he resigned, Benjamin Peirce began his professorship in mathematics and natural philosophy. Peirce's calls for reform—as loud and persistent as those of Jacob Bigelow's—would be dampened however by two events: the release of the Yale Report in 1828 that tamped down most calls for reform for several decades following its publication, and the appointment of Daniel Treadwell in 1834 as Harvard's second Rumford

Professor of Applied Science. Although a capable scientist and teacher, Treadwell was not an educational reformer. He had little interest in Bigelow's politically charged and progressive agenda of elevating science education and research within the structure of Harvard's implacable classical liberal arts curriculum.

It was not that nineteenth-century Harvard was without the means to teach science: it boasted the largest library in the nation, sponsored several professorships in mathematics and the natural sciences, held valuable collections of scientific equipment and collections of natural history specimens, maintained a botanical garden, and supported an observatory of the first rank. What it needed was a spark to light the fire of curriculum reform. The spark came soon after Edward Everett was sworn in as Harvard's eighteenth president. In 1845, eleven years after he assumed the position, Daniel Treadwell resigned his position as Rumford professor. The vacancy enabled Everett to fill the position, which he did two years later with German-educated Eben Horsford of New York.

CHAPTER 7 | *Bridge to the Future*

> *We need, then, a school not for boys, but for young men whose early education is completed, either in college or elsewhere, and now intend to enter upon an active life as engineers or chemists, or in general, as men of science, applying their attainments to practical purposes where they may learn what has been done at other times and in other countries, and may acquire habits of investigation and reflection, with an aptitude for observing and describing.*[1] —Abbott Lawrence

Born into a farm family on July 27, 1818, in Moscow, New York, south of Rochester, Ebenezer Norton Horsford attended several district schools before graduating from Livingston County High School in 1834. After working for a railroad survey crew, Horsford enrolled in Rensselaer Institute's engineering course, graduating in 1838 with a degree in civil engineering. After working for the New York State Geological Survey for the better part of a year, Horsford took a job teaching mathematics and natural science at a women's academy in Albany. Looking to elevate his status—and paycheck—Horsford quit teaching in 1844 on the advice of longtime friend John White Webster, who taught chemistry and geology at Harvard, and set sail for Giessen, Germany, to study agricultural chemistry with Justus von Liebig, who gained world renown following the 1840 publication of his book *Organic Chemistry in its Applications to Agriculture and Physiology.*

If you wanted the highest level of graduate study in chemistry, Liebig's laboratory in Giessen, Germany, was the place to be, and it was for many Americans who studied with Liebig in the late 1840s. It enabled Horsford to find immediate work with the Boston Water Works upon his return to the United States in 1846 and, one year later, to be selected as Daniel Treadwell's replacement as Rumford Professor of Applied Science, a position he would retain for the next sixteen years. The confluence of Everett, Peirce, and Horsford made the timing right to advocate for a separate school of science. Everett's vision, however, went far beyond a school of science; his was an expansive vision that included two major strands: one humanistic, the other scientific. It was an idea that he had been mulling over for quite some time.

In November of 1846, six months after Everett's inaugural address, members of Harvard Corporation voted to form a committee to report upon "a plan of a School of Science and Literature, to be established as a separate department of the University."[2] In February, members of the Corporation approved a provisional plan drawn up by Everett and submitted to them the previous month. In March, the Corporation voted to include Harvard College's professors of Latin and Greek literature to the faculty of the school, which had consisted only of scientists. When the faculty of the new school met the following month for the first time, Everett laid out his vision for a school with two distinct departments: one in the humanities, the other in the natural sciences. He followed this up in writing with a draft document titled "Programme of the Scientific School for the Academic Year 1847-48."[3] Underscoring the experimental nature of the proposed scientific school, Everett pressed his case, outlining in great detail his vision for the two departments: a humanities department that would include Greek, Latin, history, and modern languages; and a scientific department that would include astronomy, mineralogy, geology, mathematics, botany, natural philosophy, anatomy and physiology, and chemistry.

By the following summer, Everett's plan—and by then, it *was* Everett's plan—was accepted by the school's faculty, though not enthusiastically by everyone. When an outside donor stepped forward to offer a large sum of money to underwrite the experiment, there was no stopping the proposal's momentum. The outside donor was Abbott Lawrence, a Boston import merchant and textile manufacturer, who published his intended $50,000 gift and the stipulations that came with it in several Boston newspapers, citing a letter Lawrence had written to Harvard's treasurer Samuel Eliot dated June 7, 1847. On the surface, the $50,000 donation appeared to be the creative genius of Lawrence, but looking deeper into the matter, it is apparent that the idea was not his alone; or at least the ascription of credit seems to shift depending upon the source. For instance, although Everett laid out the plan for a scientific school in his 1846 inaugural address, he did so borrowing portions of Benjamin Peirce's earlier ideas. A decade later, in his memorial address for Abbott Lawrence, Everett (who didn't have the best relationship with Peirce by then) gave Lawrence full credit for the idea. At the same time, Lawrence gave Eben Horsford, the newly appointed Rumford professor, credit for the idea in the letter that he wrote to Eliot first outlining his gift. Circular? *Yes.* Confusing? *Absolutely.* But in the end, what matters is that Everett, members of the Corporation, and Harvard's Board of Overseers eagerly accepted Lawrence's gift of $50,000 for a new professional school in the applied sciences.

Like Abiel Chandler, who would place restrictions on his $50,000 gift to establish a scientific school at Dartmouth five years later, Abbott Lawrence outlined in great detail not only how much money he was willing to give to seed the scientific school, but also how he expected the gift to be used. These are revealed in the following excerpts taken from Lawrence's letter that was published in local newspapers in 1847 after the gift was announced, in the *American Journal of Education* several years after that, and in Hamilton Andrews Hill's memoir of Abbott Lawrence published in 1884. The letter, addressed to Samuel Eliot, begins:

> My Dear Sir—I have more than once conversed with you upon the subject of establishing a school for the purpose of teaching the practical sciences, in this city or neighborhood; and was gratified when I learned from you that the government of Harvard University had determined to establish such a school in Cambridge, and that a Professor had been appointed who is eminent in the science of Chemistry, and who is to be supported on the foundation created by the munificence of the late Count Rumford.[4]

Of course Lawrence was referring to the appointment of Eben Horsford as Harvard's third Rumford Professor of Applied Science. Along with chemistry, Lawrence proposed that instruction in the new school of science should be offered in geology, mineralogy, natural philosophy, engineering, and the natural sciences, emphasizing the hiring of teachers "who have practiced and are practicing the arts they are called to teach."[5] And, like its counterpart at Yale, the new school should depend upon private donations and student fees for its fiscal support: "it is the best guarantee to exertion and fidelity, and the permanent prosperity of the institution."[6]

Having outlined the generalities of his gift, Lawrence launched into a detailed description of how the gift should be spent:

> I therefore propose to offer, through you, for the acceptance of the President and Fellows of Harvard College, the sum of fifty thousand dollars, to be appropriated as I have indicated in the foregoing remarks. The buildings, I have supposed, without having made estimates, could be erected, including an extensive laboratory, for about thirty thousand dollars. If so, there will remain the sum of twenty thousand dollars; and I suggest that whatever sum may remain, after the erection and furnishing of the buildings, should form the basis of a fund, which together with one half of the tuition fees, till the amount shall yield the sum of three thousand dollars annually, shall be equally divided between the Professor of Engineering and the Professor of Geology, and be made a permanent foundation for these Professorships. The object is, to place the three Professors [the third being the Rumford Professor of Applied Science] in this School in the same pecuniary situations.[7]

Lawrence concluded the above with a plea: "I beg to suggest, further, that the whole income of this School be devoted to the acquisition, illustration, and dissemination of the practical sciences forever."[8] Although a surprisingly large donation, Lawrence's gift was not out of the ordinary for the Lawrence brothers—Abbott, Amos, and Samuel. Guided by the notion of Christian charity, the Lawrence brothers epitomized the ideal of successful businessmen giving to organizations—civic and otherwise—for the greater good of all. An expression of this ideal is found nowhere so eloquently than on a loose slip of paper that Amos Lawrence, Abbott's older brother, carried around in his wallet. On it, Amos had scribbled the biblical meditation: "What shall it profit a man, if he gain the whole world, and lose his own soul?"[9] It was a tangible reminder of his duties to help the less fortunate, duties that made Amos Lawrence a major benefactor of Williams College and Abbott Lawrence a major benefactor of Harvard's scientific school.

No matter how the scientific school came into existence, it was Abbott Lawrence's extraordinary gift to Harvard in the summer of 1847 that turned the idea for a scienific school into a reality. President Everett, members of the Corporation, and Harvard's Board of Overseers officially recognized Lawrence's gift during the 1848 commencement ceremonies when they christened the new endeavor the Lawrence Scientific School, which, of course, was headquartered in Lawrence Hall on Kirkland Street.

ALTHOUGH THE DOORS TO the Lawrence Scientific School opened in the fall of 1848 right after the institution's official christening, it would be another fourteen years before Edward Pickering would arrive at the scientific school's front door seeking a bachelor of science degree. By then several significant developments had occurred in higher education pertaining to education in the arts and sciences: New Hampshire native Abiel Chandler founded the Abiel Chandler School of Science and the Arts at Dartmouth (1852), inventor and manufacturer Peter Cooper opened Cooper Union for the Advancement of Science and Art in New York City (1859), and the Commonwealth of Massachusetts chartered the Massachusetts Institute of Technology (1861). But other than the latter development, Edward Pickering probably didn't have these developments in mind when, at the age of sixteen—a man by his era's standard—the young scholar packed his bags one autumn day, said good-bye to his family on Mount Vernon Street, and headed off to Cambridge to start a new chapter of his life at the Lawrence Scientific School.

To get an idea of both the transportation systems available to him and what sights he might see on his way to Cambridge, let's accompany Edward Pickering on his trek from his residence on Mount Vernon Street in Beacon Hill to Lawrence Hall on Harvard's campus in Cambridge. More than likely, since his family was of modest means despite their Mount Vernon Street address, Edward began by walking west along Mount Vernon Street past the southern edge of Louisburg Square to tree-lined Cedar Street.[10] There he would turn right and head north. At the end of the first block, he would cross Pinckney Street, the unspoken boundary line between Beacon Hill's fashionable South Slope neighborhood and its lower-class immigrant North Slope neighborhood (which, in 1862, would have been populated primarily by African-American families, many of whom worked in the mansions to the south). On his left, one block over, Charles Street paralleled the river, marking another boundary line: this one separating Beacon Hill's North and South Slope neighborhoods from the blue-collar, artisan-filled Flat-of-the-Hill district nestled next to the Charles River.

After crossing Pinckney, May, and Southac Streets, where he might see placards announcing an anti-slavery meeting at the African Meeting House on Smith Court or a newsboy hawking the latest issue of William Lloyd Garrison's abolitionist newspaper *The Liberator*, Pickering would arrive at the southern wall of the Charles Street Jail. Built between 1848 and 1851 at the corner of Charles and Cambridge Streets, the layout of the massive structure was both functional and symbolic: designed by Boston architect Gridley Bryant in the shape of a cross with four wings extending from a central, octagonal rotunda, the structure reminded inmates of the central Christian values of mercy and forgiveness while enabling jailers to segregate the inmate population by gender and type of offence through the convenient placement of its separate wings.

Standing in front of the thick granite walls of the jail, Pickering would wait to catch the westbound Cambridge Horse Railroad, perhaps picking up a copy of the *Boston Evening Traveller* to read as he journeyed to Cambridge. A late September issue would contain an article of interest for many Northerners: Lincoln's warning to slave states in the South that if they didn't rejoin the Union by the end of the year, he would issue an Emancipation Proclamation that would free all slaves in the Confederacy. With or without a newspaper in hand, Pickering would board one of the Cambridge Horse Railroad's streetcars that ran several times an hour. It would take him across the West Boston Bridge into East Cambridge along Main Street, a thin strip of land with the Charles River on his left and a drainage canal on his right.

Main Street ran along the Charles River for a good quarter of a mile until it merged with Front Street and angled inland, away from the river and the back-filled marshlands along its banks. The streetcar would make several stops, especially at Central Square, where the line ended in 1856 when the tracks were first laid, and then continue westward along Main Street, making stops at Bigelow, Clinton, Lee, Hancock, Dana, and Ellery Streets before turning northwest at the junction of Mount Auburn Street where many travelers would disembark and either walk or take an omnibus to the cemetery for an afternoon outing.

By 1862, Mount Auburn Cemetery was drawing thousands of visitors a year, more than Niagara Falls and Mount Vernon combined. The original seventy acres of rolling hills, winding paths, ornate gardens, funerary art, grand monuments, and sparkling lakes was the first rural cemetery in the United States, breaking with the stark burial grounds of the colonial period and dour church-affiliated graveyards. Dedicated in 1831, Mount Auburn Cemetery was both a pragmatic response to Boston's overcrowded graveyards and an aesthetic response to changing attitudes about death. The term "cemetery" itself is derived from the Greek for "a sleeping place," which reflects a more accepting attitude toward death. Conceived by Jacob Bigelow and Alexander Wadsworth, and designed by Henry Alexander Dearborn, Mount Auburn Cemetery is the final resting place of many prominent Bostonians, including Nathaniel Bowditch, Charles Bulfinch, William Ellery Channing, Henry Wadsworth Longfellow, Amy Lowell, Harrison Gray Otis, Joseph Story, and Charles Sumner.

Pickering would not disembark at the cemetery, however; he would stay aboard the horse-drawn streetcar for several more blocks until it stopped at Harvard Square, where Harvard and Brattle Streets merged at the southwest corner of Harvard Yard. After disembarking, he'd walk the rest of the way to Lawrence Hall, taking the more direct route through Harvard Yard. More than likely, he entered the Yard near a two-story, wood-frame house built in the early 1700s for Benjamin Wadsworth, Harvard's eighth president. The house was used as the president's residence until Jared Sparks, Harvard's nineteenth president (and Edward Pickering's future father-in-law), decided to live in his own house at 48 Quincy Street, at which time Wadsworth House, as it became known, was converted into a student dormitory.

From Wadsworth House, Pickering would walk north, passing Dana House on his left, built in 1832 (and moved brick-by-brick forty years later when Matthews Hall was erected in its place). The two-story neoclassical structure sported gleaming white Ionic columns and a low-slung roof with handsome recessed pediments. Named after Nathan Dane, a prominent lawyer and statesman who endowed the Dane Professorship in Law, the building was the early location of the Harvard Law School, often referred to in its early days as Dane Law School.

After passing Dane Hall, Pickering would briefly skirt the eastern edge of Massachusetts Hall. The three-story brick structure, built between 1718-1720 as a dormitory, was divided into thirty-two sleeping chambers and sixty-four private study carrels, one for each of the sixty-four students housed in the building. It remained that way until 1869, four years after Pickering graduated from the scientific school, when it was redesigned to hold lecture halls, examination rooms, and laboratories. The most distinctive feature of Massachusetts Hall, one that Pickering would not see as he walked past the eastern end of the building, was the clock face in the western gable added in 1725.

The next building Pickering passed was Old Stoughton, set at a right angle to Massachusetts Hall. But he would pass the backside of Old Stoughton, missing the quadrangle created by the placement of Massachusetts, Old Stoughton, and Harvard Hall (the latter set at a right angle north of Old Stoughton). The group of buildings, with less than five feet of space between the corners of each one, oriented westward toward Cambridge Common (which explains the clock face in Massachusetts Hall's western gable), and not toward the more open spaces—the old cow-watch yard—to the east. Although in Pickering's time, the arrangement of buildings in a quadrangle was more accidental than planned, it was the effect—unconscious or not—of trying to replicate the look and feel of England's Oxford and Cambridge universities. In *Harvard: An Architectural History*, Bainbridge Bunting notes that the apparent harmony established in the Old Yard was less from "a purist allegiance to a predetermined idiom," and more from "a family resemblance attained through the use of brick, simple massing, and a restrained familiar architectural vocabulary."[11]

Whether Edward Pickering contemplated the arrangement of buildings into rectilinear patterns or not, he must have felt some sense of awe as he passed the eastern end of Harvard Hall, the third iteration of the original hall completed in 1644. The original two-story wood-frame structure lasted little more than forty years. Filled with wood rot, it was demolished to make way for its successor—Harvard Hall II. Whereas Harvard Hall I centered around an all-purpose main room used for lectures, recitations, prayers, dinning activities, and various college functions, Harvard Hall II, contained a variety of rooms—lecture and recitation rooms, sleeping chambers, kitchen and pantry, and the college library. Unfortunately, as we learned in an earlier chapter, Harvard Hall II went up in flames in 1674 when a fire swept through the structure destroying the interior and all its contents. From its ashes, however, rose

Harvard Hall III, a beautifully designed, Georgian-style brick structure whose interior contained carved woodwork, brocade-covered walls, and an occasional splash of flock wallpaper. This is the building that Edward Pickering would have passed as he strode through the Yard in the fall of 1862 on his way to the Lawrence Scientific School.

On the immediate north side of Harvard Hall, Pickering would have stumbled upon another structure of note—Hollis Hall, built one year before Harvard Hall burned to the ground. Named after the Hollis family of London, the four-story brick building reflects the Georgian architecture that was still ascendant at the time. Built by master craftsman Thomas Dawes, the structure completed another quadrangle in the Yard, with Harvard and Hollis Halls and Holden Chapel, built in 1744, all set at right angles to each other in the same manner and orientation—westward—that Massachusetts, Old Stoughton, and Harvard Halls were to the south.

Again, whether or not young Edward Pickering chose to meditate on the finer architectural details of the Yard is uncertain. He might have chosen instead to reflect on the munificence of the Hollis family. Thomas Hollis the elder was a wealthy London merchant who held deep views about religious freedom, forged from his Baptist background that was often at odds with the Church of England. To support the liberal spirit that was gaining strength in Boston and Cambridge (and to dampen the growing intolerance of New England's Congregationalist Church), Hollis funded Harvard's first endowed chair in 1721, the Hollis Professorship of Divinity. Six years later he endowed the Hollis Professorship of Mathematics and Natural Philosophy, which did more to elevate the status of science—applied or not—than anything else during Harvard's early years.

Thomas Hollis's son also contributed heavily toward Harvard's coffers, but in a slightly different manner. An only child who inherited two family fortunes while still a teenager, Thomas Hollis the younger was an eccentric who lived most of his life in London at Lincoln's Inn, the most prestigious of the four Inns of Court of London. In his short lifetime (he died at the age of fifty-two), Hollis dedicated most of his time and energy—when he wasn't traveling, that is—to the study of the classics, modern languages, and law. After the Great Harvard Hall Fire of 1674, Hollis, a bibliophile at heart, spent the last ten years of his life shipping thousands of books across the Atlantic to Harvard's grateful but often bewildered librarians, often inserting his own bookplates into the donations, proclaiming that the benefactor was an Englishman, a Member of Lincoln's Inn of London, and a Fellow of the Society of Antiquaries.[12]

However, I doubt that Edward Pickering was thinking about Thomas Hollis the younger—or his father—on his stroll through Harvard Yard. More than likely, he was thinking about "Rebellion Tree," a stately elm in front of Hollis Hall that served as a rallying point for many a student rebellion. Even more likely, Pickering was thinking about his destination—Lawrence Hall—just beyond the northern edge of the Yard. The shortest route to Lawrence Hall would be for Pickering to take a diagonal from the southeast corner of Hollis Hall to the eastern edge of Holworthy Hall, a four-story brick building built on the eve of the War of 1812. The building, another student dormitory, was named after English merchant Matthew Holworthy, who bequeathed a considerable portion of his estate to Harvard after his death in 1678; to many, it was a sum that surpassed even that of its original benefactor John Harvard and the Hollis family. From the north side of Holworthy Hall, Pickering would get the first glimpse of his objective—Lawrence Hall, the eponymous center of Harvard's scientific school.

But Lawrence Hall probably wasn't Edward Pickering's first stop after he stepped off the Cambridge Horse Railroad at the southwestern edge of Harvard Yard. Unlike other first-year students at Harvard enrolled in the academic or classical program, Pickering would have to find his own room and board off campus as it wasn't provided by the school. As such, Pickering first would have found his lodging house, dropped his bag in his room, chatted with the proprietor for a minute or two, and then walked through the Yard to Lawrence Hall. In any case, at some point young Edward Pickering would have stood in front of Lawrence Hall, fourteen years after the school had been dedicated and its first students began to arrive.

Located on the north side of Kirkland Street (named like so many streets in Cambridge after a prominent citizen; in this case, John Thornton Kirkland), Lawrence Hall was an imposing two-story brick building—four if you count the third-floor attic space and cramped but usable dirt-floor basement. A long, sloping set of stone steps led from the sidewalk to a rapidly ascending stone stoop that ended at the front door, which was a set of center-opening double doors made of heavy wood topped by a faux fanlight. Twin pilasters rising from the stoop capped with a horizontal limestone lintel created a heavy but smart-looking entranceway, made even smarter by two vertical windows with horizontal lintels on either side of the doorway and stone quoins at the corners. The second floor picked up these architectural details sans the horizontal lintels (replaced by arched lintels) and the doorway (replaced by a third window). Everything culminated at the gabled roof that was punctuated with an ornamental circular window facing Kirkland Street. Not as bold as University Hall with its glistening white granite walls, Lawrence Hall, nonetheless, stood in relief to the heavier and more massive redbrick buildings populating most of Harvard Yard.

An engraving completed in 1849, one year after the school opened, shows the protruding main building and a smaller east wing with a separate entrance off Kirkland Street. Although diminutive in size compared to its western cousin, the east wing displayed many of the same architectural details. But all of this would change in 1871, when Lawrence Hall underwent a radical renovation in order to accommodate a larger chemistry laboratory on the first floor and additional engineering classrooms on the second and third floors. The overly ornate and massive front entrance was removed and a covered passageway between the main building and the smaller east wing was inserted, inviting guests into the building with doorways on the left leading to classrooms and laboratories and on the right leading to faculty offices and more classrooms. Another radical change involved the third-floor attic space. To accommodate a growing engineering program, the roof was raised and windows were inserted on all four sides. Twenty years later, the widow of Benjamin Rotch, founder of the New Bedford Cordage Company, would give $10,000 for the construction of a north wing that would house the electrical engineering program. A west wing was proposed, but never built, where a small, freestanding building stood that served as faculty offices until it was demolished in 1878.

Originally, Lawrence Hall was built to house the scientific school's chemistry department with classrooms, laboratories, and a small library in the main building, and Eben Horsford's residence in the smaller east wing. By 1853, the school offered courses in civil, mechanical, and electrical engineering in several upper-floor rooms under the direction of Henry Eustis, a graduate of Harvard College and the United States Military Academy at West Point. By the 1871 renovation, Eustis and his students had completely taken over the second and third floors.

According to the *List of Students of the Lawrence Scientific School, 1847-1894*, published in 1898, Edward Pickering attended the Lawrence Scientific School from the fall of 1862 to the spring of 1865. Pickering enrolled in the scientific school after his somewhat undistinguished student days at Boston Latin at the suggestion of Charles Eliot, who at the time was an assistant professor of mathematics and chemistry at Harvard and a personal acquaintance of Francis Gardner, headmaster of Boston Latin. It wasn't that Pickering was intellectually dull; it was just that the classical Latin grammar curriculum he studied at Boston Latin didn't awaken his senses, at least not in the way that the practical science classes at Lawrence Scientific School would. This was precisely the argument that educational reformers were making: the classical curriculum just wasn't for everyone; some students—many students—needed more. They needed a practical, hands-on laboratory experience where they could explore solutions to real-world problems. And that was precisely what the Lawrence Scientific School intended to provide.

When Pickering arrived at the scientific school, he entered as a first-year student in chemistry, and he must have excelled at his studies, as he was offered a position as assistant instructor at the end of the year. Once again, Charles Eliot intervened, advising the young scholar not only to decline the offer, but also to change his major to physics. Was the change to physics on the advice of Eliot alone or was there another reason that led Pickering to change his major? Without the availability of a personal diary or any other substantiating documentation, it's hard to say. One factor that might have influenced Pickering's decision, however, was the resignation of Eben Horsford at the end of Pickering's first year. Curiously, Horsford's decision was based less on his sixteen-year affiliation with Lawrence Scientific School and more on his increasingly lucrative chemistry business outside of it.[13] In any case, by his sophomore year, Edward Pickering was spending most of his time studying physics, which he did in Harvard Hall discussing readings, listening to lectures, and watching demonstrations.

STANLEY GURALNICK REMINDS US in *Science and the Ante-Bellum American College* that it is impossible to discuss physics in the nineteenth century without first discussing the terminology used to describe the field of natural philosophy of which physics was a small part. For instance, a department of mathematics and natural philosophy (which was a typical department designation) could include such disparate subjects as mathematics, astronomy, physics, chemistry, meteorology, geology, and mineralogy. In other words, "natural philosophy" was a catch-all term that had little to do with "philosophy" as we know it today, but much to do with the emerging physical sciences that didn't quite fit into the established disciplines of the classical liberal arts curriculum. It also had a lot to do with the faculty: *who* taught often determined *what* was taught, and since many of the instructors who taught natural philosophy at antebellum colleges were often tutors or lecturers, the range of what was taught varied tremendously.

Aside from an instructor's interest and professional preparation, the selection of a textbook was the second most important factor in determining course content. In the diverse field of natural philosophy, one text dominated all others: William Enfield's *Institutes of Natural Philosophy*. For want of another textbook, Enfield's book, a compilation from various sources—many of them outdated, some of them dubious—dominated the teaching of natural philosophy well into the 1820s, at which point more specialized textbooks that treated the various physical sciences as separate fields of inquiry with their own native principles and, even more importantly, mathematical bases began to appear.

In a span of less than fifty years, between 1825 and the outbreak of the American Civil War, Guralnick identifies three phases in the evolution of teaching natural philosophy as antebellum colleges tried to break from the exclusive use of Enfield's flawed text:

> The first is the period from 1825-1833 during which Enfield was replaced at all colleges by three newer American textbooks and during which separate texts on astronomy and on mechanics also appeared. The second period, 1833-1850, is marked by the use of a separate textbook on optics and the extension of teaching by "experimental lecture," especially lectures on electricity, magnetism, meteorology, and the steam engine. The third period, dating after 1850, is that in which the textbooks finally presented mechanics and contemporary electricity and magnetism in sufficient mathematical detail.[14]

This is a relevant discussion for understanding Edward Pickering's experience at Lawrence Scientific School, since he attended the school during the third period of Guralnick's assessment above. By 1862, Pickering's first year at Lawrence Scientific, our young subject would have been fully immersed in readings that were much more detailed and diversified than those who came before him, readings that treated what were once subtopics as independent areas of research, e.g., electricity, magnetism, optics, and thermodynamics. Unfortunately, when it comes to instruction, the lecture-demonstration (or, as Guralnick calls it, the "experimental lecture") was the dominant form of instruction in the sciences at Harvard (other, more "classical" subjects still relied on daily recitations). Pickering's physics instructor, Joseph Lovering, relied on the lecture-demonstration extensively. In fact, how Lovering taught physics at Harvard was no different from how Benjamin Silliman, Sr. taught chemistry at Yale. Though both men were admired for the expansiveness of their knowledge, their intellectual curiosity, and their ability to thoroughly engage an audience, they relied primarily on the tried-and-true format of the lecture-demonstration to impart knowledge to the next generation of scholars.

Lovering, however, never imagined that he would be teaching physics at Harvard at all: instead, he envisioned a future in the world of ministry. After receiving his bachelor's degree from Harvard College in 1833 (after which he taught briefly in a small private school in nearby Charlestown), Lovering entered Harvard's Divinity School, but he soon found himself working as a math tutor under Benjamin Peirce in the Department of Mathematics and Natural Philosophy. When John Farrar, Hollis Professor of Mathematics and Natural Philosophy, fell ill, Lovering took over the ailing professor's classes. Upon graduation from Divinity School in 1836, in which he gave the valedictory address, Lovering assumed the position of tutor in mathematics and lecturer in natural philosophy. Two years later, Lovering replaced the still-ailing Farrar as Harvard's sixth Hollis Professor of Mathematics and Natural Philosophy, a position he would hold for the next fifty years.

Lovering was known not only within the halls of Harvard University, but outside of them as well. Like his counterpart at Yale, Benjamin Silliman, Sr., Lovering lectured widely, appearing regularly at the Lowell Institute in Boston, the Smithsonian Institution in Washington, D.C., and the Peabody Institute in Baltimore, to name just a few of the institutions to which he was invited. Through his affiliation with William Bond at the Harvard College Observatory, he participated in the Royal Society of London's project on the earth's magnetism; and, later, he worked with Alexander Bache on the United States Coast Survey, where he was in charge of computations for determining differences of longitude in the United States.

Although reserved in his demeanor, Lovering was active socially and professionally. He was an active member of the American Academy of Arts and Sciences and the American Association for the Advancement of Science; he also held memberships at the American Philosophical Society of Philadelphia, the California Academy of Sciences, the Buffalo Historical Society, the Scientific Club of Cambridge, and the Thursday Evening Club in Boston. In addition to this, Lovering was a trustee of the Tyndall Fund for the endowment of research in physics and, during the last few years of his life, a trustee of the Peabody Museum of American Archeology and Ethnology.

But it is probably in his many diverse publications that we get a sense of the range of topics that Lovering explored throughout his long academic career, primarily as a single author, but also in collaboration with others (most notably, William Bond and Benjamin Peirce). Although most of the topics he wrote about related to astronomy, magnetism, and meteorology (e.g., Encke's Comet, the earth's magnetism, lightning and lightning-rods, the Aurora Borealis, sunspots and solar eclipses, and the nature of halos and coronae), Lovering delved into some rather unusual topics as well (e.g., the directional flow of the Mississippi River, the nature of the Australian boomerang, and the origin of the French Republican Calendar).

Despite his many achievements, Lovering represented the past. Although he was a born teacher, Joseph Lovering was not a born investigator. At a memorial sponsored by the American Academy of Arts and Science, held on February 10, 1892, Andrew Preston Peabody, the Academy's vice-president and opening eulogist, acknowledged as much when he stated:

> He was not, as he himself would be the first to avow, a born investigator, although, as his successor will doubtless tell you, he did very substantial work as an original student; but he was a great teacher, and I am persuaded that the experiences of his education to which I have referred had an important influence on the result. He came to the study of physics as a ripe literary scholar; and he dwelt on its various

fields, with their intricate relations, until he had acquired clear conceptions of the whole ground, and it was thus that he gained the power of presenting all the details with such clearness and force.[15]

Although others noted this distinction—Lovering as a great teacher—as Peabody continued his eulogy, he put his finger on one of the great tensions of the day: Is it enough for the university scientist to be a great teacher, or should he also be an excellent researcher? Here's how Peabody framed the question:

> So called "original research" is now a fad in education, and we are in danger of overlooking the fact that the scholar and teacher is no less important to the community than the investigator. It is absurd to contend which is the more important member of the body politic. No one has pressed on this community more persistently than myself the importance of scientific investigation, not primarily for the results it may yield, but chiefly as a great influence towards sustaining the higher life of the nation....[16]

Although the fuller context of the memorial bespeaks of a man who was steeped in tradition, according to others who spoke at the memorial, Lovering often held his traditional training and views at bay in order to fully understand the challenges that he faced, challenges that were of a twofold nature: on the one hand, the challenges were structural (how to organize and staff the Department of Mathematics and Natural Philosophy), and just as often they were pedagogical (what courses should be offered and how should they be taught).

Regarding the organizational challenges, in 1833, the year that Lovering graduated with a B.A. from Harvard College, the Department of Mathematics and Natural Philosophy was one unit unofficially headed by Benjamin Peirce. In 1836, Peirce was named University Professor of Mathematics and Natural Philosophy and head of the department by the same name (while John Farrar held the position of Hollis Professor of

Mathematics and Natural Philosophy). Lovering, now a tutor in mathematics and lecturer in natural philosophy, was named Peirce's assistant. A year later, when the Department of Mathematics and Natural Philosophy split into separate units, Peirce was named head of the Department of Mathematics and Lovering, with Farrar in declining health, took over the Department of Natural Philosophy. The following year, with Farrar's health still in decline, Lovering was appointed Hollis Professor of Mathematics and Natural Philosophy. Two years later, in 1840, the Department of Natural Philosophy was renamed the Department of Physics and, in 1842, to give Peirce a title of distinction, which he thoroughly deserved, Peirce was named Perkins Professor of Mathematics and Astronomy.[17]

In less than ten years, the Department of Mathematics and Natural Philosophy saw significant changes not only to its structure, but also to its personnel. But pedagogically, especially in the Department of Physics, where Lovering commanded both the curriculum and most of the teaching load, not much had changed. The lecture-demonstration format persisted, even after the establishment of the Lawrence Scientific School in 1847. As Florian Cajori notes, before the laboratory approach became prevalent in the latter half of the nineteenth century, "hard thinking" was regarded as the sole requisite for scientific discovery.[18] Hard thinking was also the fundament upon which instruction in the sciences was built, culminating in the lecture-demonstration that characterized most lyceum lecture programs and antebellum college classrooms. But as pressure mounted to provide instruction in the application of science to real-world problems, a new form of instruction was necessary, a hands-on, laboratory approach found in most European universities. It did come, of course, just not to the teaching of physics; it came first to the field of chemistry.

At Harvard, it came at the insistence of Josiah Cooke. Unlike his counterpart, Joseph Lovering, Cooke felt entirely hamstrung by the tradition of the lecture-demonstration, especially when it came to teaching chemistry. Largely self-taught (except for an eight-month foray to

Europe to study with some of the best chemists in the world), Cooke introduced laboratory teaching at Harvard shortly after his appointment as Erving Professor of Chemistry and Mineralogy in 1850, setting up a small chemistry lab in a cellar in the north end of University Hall. When Boylston Hall opened in 1858, Cooke moved the chemistry lab to the east side of the first floor of the new building. By the time of Cooke's death in 1894, the chemistry department had overtaken most of the building. Although it was a series of lectures given by Yale's Benjamin Silliman, Sr. that motivated Cooke to become a chemist, it was Cooke's experience as a teenager setting up a laboratory in his family's residence that convinced him of the importance of the experimental or laboratory method.[19] Cooke, however, was no island unto himself: he repeatedly fought to introduce science into the academic college course of study on an equal basis with the humanities and classics.

Although a member of the newly formed Lawrence Scientific School, Cooke, like Lovering, had little influence. The rising star among the faculty was Swiss-born and German-educated Louis Agassiz, who had established his reputation in Europe prior to joining the faculty at Harvard in 1847 through his radical theories about glaciation. It did not take long for the charismatic and prodigious scientist to usurp the mission of the Lawrence Scientific School, turning it away from Abbott Lawrence's original vision of a school of applied science that emphasized chemistry and engineering to a program of pure science with interests in geology, mineralogy, and zoology.[20] In either case, at the heart of the new endeavor was the fully outfitted laboratory, an idea borrowed from laboratories established in many European universities. The European laboratory would influence other institutions of higher learning in America, but no more so than at Harvard and, later, at its nearby competitor, the Massachusetts Institute of Technology.

OF THE EIGHTEEN STUDENTS from Lawrence Scientific School's graduating class of 1865, thirteen of them listed physics as their major (quite a testament to the impact of Joseph Lovering). Among these students, one name stands out: Edward Charles Pickering.[21] When he began his studies at the scientific school in 1862, the physics laboratory—if it could be called that—was a barebones affair housed in University Hall, consisting of a lecture hall, a recitation room, and a storage room for the equipment that Lovering used in his lecture-demonstrations. In 1866, Lovering moved the department and its equipment from the narrow confines of University Hall to larger quarters in Harvard Hall. A year and a half later, in the winter of 1867, Edward Pickering, an assistant professor of mathematics under Benjamin Peirce, left Harvard to teach at the newly founded Institute of Technology, now in its second year of operation and located in its new building in Boston's Back Bay. To understand this move, two other people should be brought into the discussion—William Barton Rogers and Charles William Eliot, each one responsible for bringing significant change to the nineteenth-century college curriculum, especially in the area of science and technology, and, just as importantly, each one responsible for guiding Edward Pickering in his professional career.

CHAPTER 8 | *Outshine Them All*

> *I think the time is nearly at hand for an important revolution in this whole matter of collegiate education. The old institutions with their vast funds, educating youth at enormous expense, yet fitting them for nothing truly useful or calculated to advance the age, must soon meet the rivalry of institutions which will embody modern ideas.*[1] —Henry Darwin Rogers

Writing in the middle of the twentieth century, Samuel Prescott reminds us that the careers of the Rogers brothers—four in all: James, William, Henry, and Robert—coincided with the rise of science, technology, and industry in America in the first half of the nineteenth century, with each of the brothers distinguishing himself in some branch of science while serving in a prominent academic position.[2] The impetus for their interest is explained by the adage "the apple doesn't fall far from the tree": it came unmistakably from their father, Patrick Kerr Rogers. Born in a small village in Tyrone County, Northern Ireland, the same year that America declared its independence from British rule, Patrick was an especially intelligent and curious child who received his education from the local schoolmaster and from his uncles. Rather than become a teacher or a clergyman like his uncles, Patrick decided to move to Dublin to try his hand at business.

Unfortunately, the move coincided with the Irish Rebellion of 1798, and Patrick, being the patriot that he was, published several articles against British rule in local newspapers. Believing his arrest imminent, Patrick fled Dublin with the help of kinsmen, arriving in Londonderry in mid-summer. By August, twenty-two-year-old Patrick Rogers was on a ship headed to America.

He settled in Philadelphia, the intellectual epicenter of the new republic, and within months was admitted to the medical school of the University of Pennsylvania, where he studied with some of the top scientists of his day: physician and chemist James Woodhouse, who founded one of the first professional organizations for chemists, the Chemical Society of Philadelphia; botanist Benjamin Smith Barton, author of *Fragments of the Natural History of Pennsylvania* and recipient of the nation's first chair of natural history; Benjamin Rush, chemist, physician, and educational reformer (his plan for a federal university was one of the earliest to call for significant modifications to the classical liberal arts model inherited from the English scholastic system); and William Shippen, Jr. and Caspar Wistar, two of the most respected physicians teaching in the medical school at the University of Pennsylvania.[3]

Between the start of his medical studies and his graduation in May of 1802 (with a thesis on the chemical and therapeutic properties of the tulip tree), Patrick met and married Hannah Blythe, who had immigrated to America with her sisters from Londonderry, and fathered the first of four sons, James Blythe Rogers. But starting a family and establishing a medical practice at the same time proved taxing. Matters were compounded when his father died in 1803, and Patrick, as the oldest son, returned to Ireland to settle family matters. His yearlong absence put him out of touch with his patients, making his financial situation dire. To make ends meet, Patrick prepared a series of popular lectures on chemistry that he accompanied with experimental demonstrations, a novelty in America at the time, which he followed up with a small publication titled "A Syllabus of a Course of Lectures on Natural Philosophy and Chemistry with the Applications of the Latter to Several of the Arts."[4] Although his private practice and the stipends he received from his public lectures hardly paid his mounting bills, they did give him a reputation of someone keenly interested in the practical application of science to the industrial arts.

By 1812, however, it was clear that Patrick needed to seek his fortune elsewhere (as a college town with a thriving medical school, Philadelphia was awash in newly-minted physicians). On the recommendation of several friends, and even a few creditors, Patrick moved to Baltimore, where he opened an apothecary shop and sought other ways to make money. He had to: before Patrick and Hannah moved to Baltimore, Hannah had given birth to two more sons, William Barton Rogers and Henry Darwin Rogers (and two daughters, both of whom died in infancy). The year after their arrival, Hannah gave birth to the couple's fourth and final son, Robert Empie Rogers. Although he had a growing family, Patrick was able to make ends meet through the proceeds of the apothecary shop, his small but growing private practice, and stipends received from occasional lecture programs. As his family and fortune grew, so too did Patrick's reputation: in 1816, he was elected physician of the Hibernian Society, and in 1819 a member of the Medico-Chirurgical Society of Baltimore.

The year 1819 was significant in another way: Thomas Jefferson's dream of a university focused on science rather than religion was becoming a reality. Patrick saw it as an opportunity and wrote to Jefferson directly, asking for a position as professor of natural philosophy at the recently chartered university. While the proposal was being considered (but ultimately declined, primarily because the opening of Jefferson's brainchild was postponed), Patrick received an offer from the College of William and Mary to teach natural history and chemistry. With an appointment to the faculty of the University of Virginia growing less and less likely, Patrick accepted William and Mary's offer and moved to Williamsburg, Virginia, settling in Brafferton House with his family in the fall of 1819.[5] Patrick reveled in the change in location, his new position, and his stable income—until catastrophe struck. The following summer Hannah contracted malaria and died. Patrick was devastated, both by the loss of his wife and the loss of his sons' mother. Like many who have suffered great loss, Patrick threw himself into his work. But he also did something else: he involved his sons in it as well, in all aspects of it, but especially in the construction of equipment for his classroom demonstrations. For the next few years, father and sons lived and worked together, forming a close-knit kinship.

IN 1825, AFTER COMPLETING their studies at the College of William and Mary, William and Henry Rogers left Williamsburg and headed for Baltimore, where James, the older brother, was enrolled in the University of Maryland's medical school. Their plan was to open a private Latin grammar school in nearby Windsor, Maryland. They did so, enrolling their youngest brother, Robert, in the newly established academy. This meant that the Rogers brothers were united once again. But the school didn't turn out to be that profitable, at least profitable enough to support two schoolmasters, so William left to seek other work. He found it as a lecturer at the newly opened Maryland Institute, where he offered courses in mathematics, physics, chemistry, and astronomy.[6] As early as 1827, William was thinking about the practical applications of science, as the following excerpt from a lecture given at the Maryland Institute on January 15, 1827, reveals:

> I need not in this place enlarge upon the usefulness of popular courses of scientific instruction; with respect to my own department, this, I hope, will be clearly evinced in another part of the present discourse. Of late years, the public mind, both in this country and abroad, has been much interested in the subject. In many places institutions calculated to render useful science attainable by the mass of society have been established; and such is the growing impression of their value that their number continues yearly to increase. Our own city has not been backward in this career of improvement. The Maryland Institute is, I believe, the second in point of seniority in the United States, and has now been upwards of a year in successful operation.[7]

Josiah Holbrook, father of the American Lyceum Movement, could have easily delivered this oration, as it is in keeping with Holbrook's emphasis on the diffusion of popular science and practical knowledge to the general population. William's performance at the Maryland Institute must have garnered the attention of the Board of Governors, as the body considered William for a fulltime position soon thereafter. Although the fulltime position fell through, William was given the opportunity to lecture at the Institute the following season, and again to great success, even though he had "little more than the blackboard and chalk."[8] Once again, his performance was rewarded, though not with a task commensurate with his performance: he was asked to draw up a plan for a technical high school, much the same as the one affiliated with the Franklin Institute in Philadelphia. In a letter to the Board of Governors dated April 13, 1828, William stated the aim of the school: "to impart such knowledge and to induce such habits of mind as may be most beneficial to youth engaging in mechanical and mercantile employments..."[9] Declaring that classical studies would not be within the scope of the school's curriculum, William went on to delineate further the school's focus:

> The earlier classes will be instructed in arithmetic, reading, writing, grammar and geography; the more advanced, in algebra, geometry, mensuration, surveying, navigation, perspective, etc., and perhaps in English composition. The latter grade of scholars, after having made a certain proficiency in their mathematical studies, will be taught the elementary principles of astronomy, mechanics, natural philosophy and chemistry, and will be permitted to attend the lectures in the Institute in aid of their scientific studies, as a reward for their diligence and improvement.[10]

After the Board of Governors accepted William's proposal, Henry closed the academy in Windsor and joined William at the technical school, as did James, who, after receiving his medical degree, was working—rather unhappily—as an industrial chemist for a manufacturer in Baltimore. Robert joined them as well, as a student at the Maryland Institute. Once again, the Rogers brothers lived and worked together as a close-knit family unit, until catastrophe struck a second time.

On his way to visit his sons in Baltimore in the summer of 1828, Patrick Rogers contracted malaria and died in Ellicott Mills, Maryland. It was the beginning of August, the College of William and Mary's fall term was around the corner, and Patrick's position was still open.

With little time to attract another candidate, the trustees of the college turned to William, who, through his lectures at the Maryland Institute, had shown a considerable grasp of scientific principles, and asked him to fill the post vacated by his father. The Rogers brothers would experience this throughout their adult lives: united and working together, then split apart but still in close contact with each other through letter correspondence.[11]

Leaving his brothers in Baltimore, William moved to Williamsburg in the fall of 1828 in time to give the opening address of the new school year at the historic college. One by one, the three remaining Rogers brothers left Baltimore. Henry left first, in 1830, after accepting a professorship in chemistry and natural philosophy at Dickinson College in Carlisle, Pennsylvania. Robert left the following year, traveling to New York with William, who served as Robert's surrogate father, in the hope of securing work on a railway survey crew (which he did, joined by Henry a year later after he resigned from Dickinson College dissatisfied with the institution's overly conservative and narrowly focused curriculum). James, the last to leave, remained in Baltimore, where he taught chemistry at Washington College until 1835, at which time he left to teach chemistry at a medical school in Cincinnati.

William's growing role as his brothers' surrogate father and mentor is further revealed in a letter to James dated March 27, 1830, as James, the last to leave Baltimore, remains teaching at Washington College, then a second-rate medical school:

> In the season of disengagement from the duties of instruction, do not abandon your studious pursuits. Do not permit your armor to rust, but keep it well burnished by continual use, and be ever ready for the field. Above all, my dear brother, be not too diffident of yourself when a favorable occasion is presented for a display of your claims to the attention of the community. This is not a country in which retiring merit is ever likely to be rewarded...[12]

William's great concern and expansiveness of spirit toward his brothers (and later his students and colleagues) is evident throughout his life, even as the four brothers began to make their own way in the world, each taking a slightly different path in the world of academia: James and Robert in the world of medicine; William and Henry in the world of science, primarily geology and mineralogy, an interest generated by Henry after a trip to England in the 1830s.

AFTER WORKING WITH ROBERT on several survey crews in the northeast, Henry quit and, as many educated men of his era did, sailed to Europe, spending the better part of a year in London, returning in the spring of 1833 with a strong interest in geology. This was due primarily to members of the Geological Society of London, who encouraged him to study the geology and topography of the surrounding countryside. This experience piqued Henry's interest, of which, in a letter to William dated January 5, 1833, he wrote about most eloquently:

> The lectures throughout London being suspended, I embraced the holidays to make a short excursion with an acquaintance into the country, to see a little of England's geology. He being, like many of the English, an excellent walker, and knowing how beneficial the exercise is always to myself, we went on foot, and shaped our rambles toward the lower end of Kent. Leaving our place of lodging in London in the evening, we walked sixteen miles, a light frost on the ground, a bright moon above, a smooth footpath leading us over hill and dale, the mists of night sleeping in the valleys, and once in every while a solitary horseman on patrol saluting us with the protecting words, "Good-night."[13]

In London, Henry attended lectures at many of London's most important scientific societies, his entry secured after he was elected a Fellow of the Geological Society of London. He listened to lectures by Dalton, Ritchie, Cummings, Turner, Babbage, Lubbock, Davies, Gilbert—all the great men of the age in science—and, the greatest of all, Faraday, the latter of which Henry had this to say: "Faraday's style of lecturing and experimenting reminds me of Paganini's playing: so easy, so adroit, *so much execution*."[14] Despite his admiration of Faraday's brilliant performance, it's what Henry said next that is most telling:

> When I listen to his fluent and eloquent delivery my thoughts wander home to you, William; and with tenderness and with a sweet pride I think of the greater powers possessed by my own dear brother. Yes, William, I have already heard several lecturers, reputed among the best in Europe, and I will vouch for it that with equal aids you shall outshine them all.[15]

Henry's experience in England resulted in two significant outcomes upon his return to America: first of all, it gave him sufficient experience to develop a series of lectures on geology that he delivered, with much acclaim, at the Franklin Institute in Philadelphia; secondly, because of his success at the Franklin Institute, he was offered a position as professor of geology at the University of Pennsylvania, where, coincidentally, after leaving the survey crew, Robert had moved to study medicine.

There was another outcome of Henry's visit to London, however: his exuberance for geology spilled over to William, who was in his fifth year of teaching at the College of William and Mary. As a result of his interactions with Henry, William turned his attention more and more toward geology, especially to the geology of the Commonwealth of Virginia. One of the first areas he began to explore was Virginia's "green sand," or marl, which he analyzed for its nutrient benefits, seeing it as an antidote to fields that had long since been depleted of nutrients from overuse. His preliminary explorations led to the publication of his first scientific paper, which appeared in Virginia's *Farmer's Register*. Subsequent papers followed, including several papers published in the *American Journal of Science*, some written by William exclusively, others with Henry as co-author. By 1835, William and Henry Rogers were two of the leading geologists in the country, and now both were teaching at prestigious institutions: Henry at the University of Pennsylvania and William at the University of Virginia, a position he had recently accepted. Their interest in geology (as well as mineralogy and topography), their growing involvement in scientific organizations at the state and national level, and their respective professorships at credible institutions led the middle brothers to accept another professional commitment: overseeing a state geological survey (Henry in New Jersey and Pennsylvania, William in Virginia).

Of all the changes in William's life, his move to the higher, cooler hill country of Charlottesville was the most pleasing, since he had suffered from the hot and humid climate in Williamsburg from an early age. Although health issues would dog him throughout his life, for now, removed from Williamsburg, embarking on a new teaching position and anticipating his involvement in the state geological survey, William's spirit was uplifted, as the following note penned to a friend in the spring of 1835 suggests:

> Spring is now exulting in the hills and valleys; graceful and lovely is the livery she wears. The soft green of the tender grass and grain that overspreads the fields and meadows; the deeper the hue of the luxuriant clover; the rich coloring of the verdure that spreads its ample folds even to the summits of the mountains,—are a delicious luxury to the eyes. Our gardens and lawn are beautiful beyond description. Just now the early roses are turning their blushing cheeks to the kisses of the sun, and the flowering locusts stand around on our lawn like bridal nymphs arrayed in white plumes and flowing lace. Odors are wafted by every breeze, and the songs of the spring birds awaken many a tender and many a sad remembrance. Surely this world is beautiful, and God is good.[16]

Three things stand out in this letter: first of all, William's keen sense of observation of the natural world around him; second, his ability to use descriptive language to capture his thoughts and impressions; and, third, his bright, upbeat countenance.

Although by 1835, the University of Virginia had strayed from Jefferson's original design—as an elective-based, science-focused institution—the faculty was more receptive to innovation than that of the College of William and Mary. It was William's new laboratory and he went to work quickly, proposing a new school that would emphasize what was beginning to emerge as his "useful arts" idea.[17] The proposal ultimately led to the founding of a school of engineering based on the blending of theory and practice. But a new school without a new approach to pedagogy was a hollow shell, and even though the school of engineering attempted to blend theory and practice, it lacked an essential ingredient—laboratory instruction. William fervently believed that "recitations, lectures, written exams, and demonstrations paled in comparison to work done in the laboratory, which allowed students to experiment with connections between theory and practice that no other form of instruction could provide."[18]

William's interest in laboratory instruction was not extracted from a textbook, but born out of his experience in the field. During the years that he managed the Virginia State Geological Survey, William experienced the repeated frustration of managing assistants who were not trained in the field; their experience, if any, came from readings and lecture-demonstrations from instructors whose knowledge had been gained precisely in the same manner. Fieldwork was barely a part of anyone's education, and if it was, it was the exception, not the rule.

During his survey years, 1835-1842, William's ideas about education related to the sciences began to gel, but he had yet to put them down on paper in a coherent manner, except in correspondence with his brothers. Aside from the prospectus he wrote for a secondary school for the Maryland Institute, William had put most of his energy into writing scientific papers related to his work as a geologist. That was about to change.

In 1837, William got a chance to work out his ideas on paper when trustees of the Franklin Institute asked him to draft a prospectus for a "School of Arts" with which to petition the Pennsylvania General Assembly. William's proposal addressed professional training for the mechanic or industrial arts in keeping with the aims and goals of the American Lyceum Movement that was sweeping the country at the time.

Unlike programs at town lyceums and mechanics' institutes that relied almost exclusively on the lecture-demonstration, William's proposal put laboratory experience at the heart of his plan. And by "laboratory," William didn't always mean an outfitted room within the walls of a brick building: the outdoors, especially for geology, engineering, and agriculture, was a more desirable place to teach the requisite observational and operational skills necessary to succeed as a geologist, a civil engineer, a miner, a surveyor, or a farmer. Unfortunately, William's proposal, titled *For the Establishment of a School of Arts, Memorial of the Franklin Institute, of the State of Pennsylvania, for the Promotion of the Mechanic Arts, to the Legislature of Pennsylvania*,[19] arrived on the desk of state legislators at the same time that the financial panic of 1837 depleted state coffers. In other words, it was dead on arrival: no legislator was in the mood to fund a new endeavor, especially one without proven merits.

The exercise, however, was not completely futile: it inspired William to write and, within the next few years, he published two books from his lecture notes: *An Elementary Treatise on the Strength of Materials*, which stressed the relationship between the practical and the theoretical, and *Elements of Mechanical Philosophy*, which was as much a treatise on physics as it was a meditation on the two branches of study related to the natural world: natural history and natural philosophy. Despite the success of William's publications, his greatest publication was yet to come: great not for its conceptual grasp of natural philosophy (at least great in the way that Darwin's *On the Origin of Species* would be), but great for William's contribution—and Henry's, his co-author—to the emerging field of geology. The paper, titled "On the Structure of the Appalachian Chain, as Exemplifying the Laws which have Regulated the Elevation of Great Mountain Chains Generally,"[20] was read in Boston in the spring of 1842 at the third annual meeting of Association of American Geologists and Naturalists (of which William and Henry were founding members), and discussed, as the title suggests, the origin of the Appalachian Mountain chain. The paper, based on the work that the brothers did on their respective state geological surveys, brought together two competing theories of mountain formation: vertical thrust and horizontal movement. Rather than seeing these ideas as competitive, William and Henry saw them as complementary, each one contributing to mountain formation by creating wave-like oscillations in the still-forming earth's crust.

What was more important, however, regarding their presentation (aside from the fact that they were elected honorary members of the Boston Society of Natural History) was the context itself—Boston. It marked William's first visit to the city, which left an indelible impression on him. For someone who had worked most of his adult life in the South, William was struck not only by the intellectual activity of the city, but also by its thriving commercial activity. It was a city that William would return to several times in the next decade, and with each visit would convince him that it was the best place to establish a school that blended the practical and the theoretical.

William's plan for a school that would offer training in the applied industrial sciences, however, would have to wait: at the start of the 1844-45 school year, the faculty of the University of Virginia elected William its chairman, a position that was equivalent to college president at other institutions. Although the yearlong position (filled with student riots, a faculty member's murder, and a battle with the state legislature that wanted to merge the institution with the state's military academy) gave William a taste of an administrative position, it postponed his plans to leave the area.[21] For several years William had been toying with the idea of moving to Boston to work on his plans for a technical institute.

At the same time that William was shepherding the University of Virginia's faculty through a tough school year, Henry, who was teaching at the University of Pennsylvania, was invited to give a series of lectures on geology at the Lowell Institute in Boston. The invitation came from John Amory Lowell, trustee of the Lowell Institute and a prominent member of the Harvard Corporation. Henry's lectures were so successful that the following year, he resigned from the University of Pennsylvania and moved to Boston, where, for a time, he was a candidate for Harvard's Rumford professorship. Failing to receive the Rumford appointment, Henry approached John Amory Lowell about creating a technical school as part of the Lowell Institute. In a letter dated March 8, 1846, Henry shared his ideas about the school with William:

> Mr. Lowell, with whom I have been talking, after mentioning the features of the Lowell will which enjoins the creation of classes in the Institute to receive exact instruction in useful knowledge, requested me to give him, in writing, the views I had just been unfolding of the value of a School of Arts as a branch of the Lowell Institute. My communication to the corporation has, I am sure, made an impression on him, and it is possible he has seen, by what is there stated, the importance of teaching science in its applied forms in this community...His plan would be to teach the operative classes of society,—builders, engineers, practical chemists, manufacturers, etc.; to admit in the first year only in limited numbers, and to teach them regularly; to have, perhaps, two permanent and salaried professors at the head of it, and to make up the rest of the instruction by assistants and by teachers, who would give courses of instruction occasionally on special branches.[22]

Henry concluded the letter by imploring William to aid him in this endeavor, speculating how great it would be if they were to head the school, working side by side—as they had for a brief time at the academy in Windsor, Maryland—in a more satisfying way than in their own professorships at separate institutions:

> How much I want you near me at this time to aid me in digesting and submitting my views on this important scheme to Mr. Lowell! If you and myself could be at the head of this Polytechnic School of the Useful Arts, it would be pleasanter for us than any college professorship, for there would be less discipline, indeed, no more than with medical students. At no distant day, if not indeed soon, Mr. Lowell will, I hope, organize such a branch in his Institute; and if he does not, you and I can surely get one founded here by going about it in the right way.[23]

Writing to his brother only five days later, William shared his conviction that "the congenial air of Boston"[24] was the best place in America for a technical school:

> Ever since I have known something of the knowledge-seeking spirit, and the intellectual capabilities of the community in and around Boston, I have felt persuaded that of all places in the world it was the one most certain to derive the highest benefits from a Polytechnic Institution. The occupations and interests of the great mass of the people are immediately connected with the applications of physical science, and their quick intelligence has already impressed them with just ideas of the value of scientific teaching in their daily pursuits.[25]

To his letter of March 13, 1846, William attached his proposal for a polytechnic school that would provide "an advanced education for the mind and thorough training of the hand."[26] The proposal, titled "A Plan for a Polytechnic School in Boston,"[27] built upon the proposal William drafted a decade earlier for the Franklin Institute. After an opening statement declaring the need for such an institution and its comprehensive nature, William defined two essential departments—physics and chemistry—around which the curriculum would revolve, but also outlined other important instructional areas: mathematics, drawing and modeling, and the modern languages of French and German, in which much of the scientific literature of the day was written. William envisioned more than the casual atmosphere of a town lyceum or mechanics' institute and more than a vocational training program geared toward turning out skilled craftsmen. William fully embraced the idea of a polytechnic school, a school that not only blended various technical domains, but also blended the theoretical and the practical as the following statement suggests:

> The true and only practicable object of a polytechnic school is, as I conceive, the teaching, not of the minute details and manipulations of the arts, which can be done only in the workshop, but the inculcation of those scientific principles

which form the basis and explanation of them, and along with this a full and methodical review of all their leading processes and operations in connection with physical laws.[28]

According to William, it was only when the mechanic, the chemist, the architect, the manufacturer, and the engineer clearly comprehended "the agencies of the materials and instruments with which he works"[29] that not only could he avoid the "disasters of blind experiment,"[30] but also make consistent and sizeable gains in his field of endeavor. After citing a multitude of examples, from the textile industry, mining and metallurgy, even the production of sugar, William concluded his "long but still incomplete catalogue of illustrations"[31] with the following cautionary note:

> We may safely affirm that there is no branch of practical industry, whether in the arts of construction, manufactures or agriculture, which is not capable of being better practiced, and even of being improved in its processes, through the knowledge of its connections with physical truths and laws, and therefore we would add that there is no class of operatives to whom the teaching of science may not become of direct and substantial utility and material usefulness.[32]

William's plan for the Lowell Institute was the most articulate expression of his idea for the establishment of a polytechnic school, going far beyond the apprenticeship, the trade school, the town lyceum, or the scientific department that a handful of colleges were beginning to consider. In the end, however, John Amory Lowell rejected William's proposal, as it fell outside the parameters of the Lowell trust. Again, the exercise was not futile: it gave William a second chance to refine his thoughts of a polytechnic institution for the industrial or mechanical arts. Although the rejection was a setback, William did not give up on the idea and, after spending the summer of 1847 in Boston, he returned to Charlottesville and the following spring announced his resignation from the university, ostensibly with the intent to move to Boston as an independent scholar and reformer. William's letter of resignation caused such a stir with trustees, faculty, and students, however, that he subsequently withdrew the letter and remained in his position, primarily out of a sense of duty to his colleagues and to the university.

Although William remained at the University of Virginia for another five years, the urge to move north, to Boston, to the Athens of America (or, as Oliver Wendell Holmes dubbed it, "the Hub of the Universe"[33]), continued to fester inside of him. Several things held him back, however: first of all, Robert, for whom he felt responsible, was teaching at the University of Virginia; secondly, the work he did for the Virginia State Geological Survey was still bearing fruit; and, thirdly, though seemingly minor, Charlottesville's temperate climate was beneficial to William's sometimes delicate health. But there were just as many reasons to move: his wife, Emma Savage Rogers, whom he married in 1846, was from Boston, and Henry, who had moved to Boston several years earlier, was becoming a fixture in Boston's scientific community. William's yearning to leave the South is expressed in a letter to Henry dated April 5, 1846, the year William drafted the Lowell Institute proposal:

> Dear Henry,—I am yearly becoming more impatient of the lifeless routine around me, and will indeed be most happy to join you in any such scheme as that of a Polytechnic Institution, either under the wing of the Lowell endowment, or other suitable and adequate auspices. You will soon begin, even in Boston, to enjoy the soft air and freshening verdure and bloom of spring. I confess, my dear Henry, I *long* to be able to walk with you and other friends on the Mall, to take an exhilarating drive to Brookline, or Cambridge, or the other neighboring points of attraction, but above all to feel the impulses of a higher social life, which have so stirred my thoughts in my visits to New England.[34]

Finally, in 1853, after eighteen years on the faculty, William resigned his position at the University of Virginia, and with Emma, his wife of seven years, headed to Boston. The motivating force was not only the strong pull of Boston's highly charged intellectual atmosphere and the hometown of his wife's family, but also the death of James Blythe Rogers, the oldest of the Rogers brothers, who had passed away the previous year. It was not so much James's death, but the fact that Robert was appointed to succeed him at the University of Pennsylvania's medical school. With Robert headed to Philadelphia and Henry already in Boston, William finally felt free to leave his friends and colleagues in Virginia.

CHAPTER 9 | *Objects and Plan*

> *Our (or rather my) Memorial is before the Legislature, and will probably be acted on next week. Thus far the sentiment is strongly in its favor. We ask for from eight to ten acres of the Back Bay land for buildings to accommodate the Natural History, Horticultural, Agricultural, Technological, etc., societies. The plan is magnificent, and if carried out will do great service.*[1] —William Barton Rogers

William and Emma Rogers spent the summer of 1853 settling into their new home, which was Emma's family residence on Temple Place, a short tree-lined street that bisected Tremont Street bordering the northeast corner of Boston Common, which they shared with Emma's recently widowed father. As summer turned into fall, William and Emma departed for Virginia so William could be close to the state legislature in order to convince legislators of the need to fund Virginia's last geological survey. The couple spent three months in Richmond, but to no avail: the legislature failed to appropriate the funds necessary for William to complete the final write-up and publication of the report. Disheartened, William and Emma returned to Boston, where William continued his outreach to Boston's civic and scientific-minded groups, trying to find a place for himself within the intellectual and cultural communities that defined Boston society at the time.

Of the two brothers, Henry was the better known in Boston. Eventually, William would find his footing among the city's intellectual and civic-minded leaders, but he would have to do so in a slow, step-by-step manner, finding work—lectureships—wherever he could. Due to Henry's growing influence, William was invited to deliver a series of lectures at the Lowell Institute on the application of science to the arts, a topic Henry had touched upon in an earlier lecture series. Over the next few years, William would give lectures at a number of scientific societies and institutes, including the Lyceum of Natural History at Williams College, the Mercantile Library Association in Boston, and the Lawrence Scientific School in Cambridge. He would also participate, often as a founding member, in the development of several scientific societies. Like Henry, William would become known as an accomplished scientist and, perhaps unique unto himself, as a brilliant orator.

In 1857, the Rogers brothers experienced another departure, but not through death: Henry sailed for Scotland to take up an appointment as Regius Professor of Natural History at the University of Glasgow. Although removed physically—separated by an ocean—William and Henry (and Robert) continued to correspond with each other, especially about their evolving ideas concerning the founding of a technical school in Boston. Although William's letters brimmed with ideas about such a school, they were full of so many other things as well, indicating his vast interests and storehouse of knowledge. He wrote about books he read, lectures he attended, people he met, and science experiments he either read about or conducted himself. He also wrote about the ever-changing state of politics, especially the question of slavery in the western territories and the impact of John Brown's raid on Harper's Ferry. But most of all, he wrote about the natural wonders around him, whether he observed them in the laboratory or in the nature itself. For instance, when Donati's Comet passed overhead in the fall of 1858, William wrote to Henry with passion and eloquence of the comet's "train" or tail:

> As seen on the night of the 2d of October, it had the aspect, when largest, of a magnificent eagle-feather, having a gracefully curving outline above, but a less regular and defined limit beneath, reaching, with its faintly vanishing end, through a superb ascending arc to near the end of the tail of the Great Bear.[2]

Aside from reflecting on the founding of a technical institute in Boston in his letters to his brothers, William had no real chance to work out his ideas until 1859. After the state legislature authorized the draining and filling of Boston's Back Bay to create new land for residential and commercial development (following Uriah Cotting's dismal failure to establish an industrial sector tapping the drainage power of the Back Bay basin), Massachusetts governor Nathaniel Banks recommended that, beyond the residential and commercial improvements envisioned, there should also be educational improvements as well: that the first charge made upon the Back Bay property "be for the enlargement of the public school fund," a fund established in 1834 by the legislature using unappropriated funds in the state coffers.[3]

Three competing proposals surfaced within a short period of time. The first, and probably the one that was most on the governor's mind, was a proposal by Louis Agassiz of Harvard's Lawrence Scientific School asking for state funds from Back Bay land sales to help with the establishment of a Museum of Comparative Zoology in Cambridge. Agassiz had already secured a commitment from Abbott Lawrence to help with instructor salaries and one from industrialist Francis Calley Gray to help with the operational side of the museum. What Agassiz lacked was money for the design and construction of the museum itself—in short, capital investment to actually break ground. Hence Agassiz's proposal to the state legislature, which entailed diverting proceeds of Back Bay land sales to his "great museum" project and, as such, away from the public school fund.[4]

While Governor Banks supported Agassiz's project, former governor and current secretary of the State Board of Education George Boutwell didn't. Boutwell was adamant that proceeds from all land sales go into the school fund to increase teacher salaries and provide greater resources for Boston's public schools. It was Boutwell's firm conviction that put him at odds not only with Agassiz's proposal, but also with a proposal by a consortium of arts and science societies led by members of the Boston Society of Natural History and the Massachusetts Horticultural Society.

The consortium, referred to as the Associated Institutions, was interested in obtaining a state land grant to establish a Conservatory of Arts and Science on four squares of land in the Back Bay, with each square devoted to one of four interests: agriculture, natural history, manufacturing, and fine arts.[5] Although William, who was away on a four-city lecture tour in Virginia, was not personally involved in creating the petition on behalf of the Associated Institutions, he was a member of the committee of the Boston Society of Natural History that spearheaded the proposal. In a letter to Henry dated February 14, 1859, William related to his brother that the plan's goal was "to induce the Legislature to set aside a large lot in the Back Bay improvement for the reception of a grand cruciform structure for the museum and libraries of the various societies and for a grand polytechnic depository."[6]

As long as Banks was governor and Boutwell was secretary of education, it was doubtful that the Associated Institutions' plan would pass the state legislature. In their detailed analysis of the origin of the Massachusetts Institute of Technology, Julius Stratton and Loretta Mannix cite Banks' "cold reception" of the conservatory committee's ideas, Boutwell's commitment to increasing the school fund, and Agassiz's persistent appeals for financial support for his museum project in Cambridge as three "interrelated deterrents" preventing passage of the Associated Institutions' proposal.[7] Of the three deterrents, the sanctity of the school fund would prove to be the greatest hurdle for the consortium's proposal over the next two legislative sessions. For the moment, however, rather than vote the Associated Institutions' proposal down, the state legislature, perhaps signaling mild support for it—or parts thereof—simply ended the 1959 legislative session without acting on it. Undaunted, members of the Associated Institutions reconstituted the Reservation Committee, the body that created the original proposal under the direction of Samuel Kneeland, inviting William Rogers to participate. It was the Reservation Committee's hope that, through his scientific interests and persuasive oratory skills, William could better formulate a petition that the 1860 legislature could accept.

Rather than draw up an entirely new petition, William used the idea behind the four-square structure designed by English-trained architect William Waud, elaborating upon each division with greater detail, emphasizing the need for interaction among the four constituent interests as a way of arousing public interest. The new emphasis notwithstanding, William's proposal was still "an outline for a collocation of educational museums, some not even in existence, which would lay out their collections for the inspection and elevation of the public."[8] Embedded within the proposal was the provision, or at least the hope, that in the not-so-distant future, courses of public lectures and demonstrations might be offered in addition to the display of objects, machinery, and other technological inventions of the day. Additionally, there was the suggestion that the educational value of the museum collections relating to the industrial arts might one day lead to the establishment of a polytechnic institute for the education of the industrial classes. It was as complete a petition William could generate given the original interests of the consortium. In a letter to Henry dated November 1, 1859, William described the efforts of the consortium to secure "a long parallelogram of the new-made land west of the Public Garden and parallel with the lower part of the Milldam," adding with a degree of optimism that "they have good hopes of succeeding."[9]

Two months later, at the start of the 1860 legislative session, William again shared his enthusiasm with Henry concerning the proposal he drafted for the Associated Institutions, indicating that the proposal had been enthusiastically received. An article in the *Boston Journal* dated February 17, 1860, recounted an address William gave to a small group of interested persons in which he addressed the general need for the study of natural science in the schools:

> Professor W. B. Rogers delivered an address on the evening of February 16, the fourth of the legislative educational meetings, on "Education as related to the Natural Sciences." He spoke with great eloquence for an hour to a small but appreciative audience, who braved the storm to secure what they knew would be a rich intellectual treat. He advocated

> the introduction of the study of natural science into the schools, not only by books alone, but by the actual objects, frequently displayed, explained and handled. As a means of strengthening the mind, cultivating the memory and the senses of observation, and of forming habits of accurate examination, there is no mental exercise superior to the study of natural science.[10]

The article goes on to describe William's proposal for the consortium of civic groups petitioning the government for land in the Back Bay, asking rhetorically:

> What student of science, miner, cultivator and owner of land, is not interested in a collection of animals, plants, minerals, ores and the various products and processes of their conversion into useful articles? What mechanic and manufacturer in the whole Commonwealth will not reap untold advantages in seeing models of implements and machinery, of the inventive genius of the world, and of perfected products of all nations, in such a Technological museum, with its explanatory lectures, as forms one department of this plan? What lover of the beautiful in art in the most remote corner of the State will not hail with pleasure the inauguration of the Fine-Art gallery which forms another feature of this plan?[11]

Despite the enthusiasm of the *Boston Journal*, and support garnered from a number of prominent institutions, including the American Academy of Arts and Sciences, the Massachusetts Charitable Mechanics Association, and the New England Society for the Promotion of Manufactures and Mechanic Arts, passage of the petition was not guaranteed. In a letter to Henry dated March 20, 1860, William acknowledged as much: "Our Memorial hangs in the Senate, where it meets opposition. It passed the House almost unanimously, and I think will succeed in the Senate. My next letter will tell the result."[12] And, indeed, William's next letter a week later did just that:

> After delays and a reconsideration the Senate have finally refused to grant the Back Bay reservation for which we applied. This result we had not dreamed of while the matter was pending in the lower House, and its great success there made us at first quite confident that it would encounter no serious opposition in the Senate. But meanwhile some enemies of the bill were quietly preoccupying the minds of the senators, so that when the time for the action drew near we found that the narrow financial views instilled into them could not be corrected.[13]

In fact, the Massachusetts lower house had endorsed two of the four divisions, forwarding the truncated petition to the Senate for review. State representatives favored the agricultural and natural history components of the petition because the Massachusetts Horticultural Society and the Boston Society of Natural History were extant institutions ready to move forward and build on Back Bay lands, whereas the manufacturing and fine arts components had no similar institutional presence in Boston. Even with the smaller scope of the petition moving forward, Boutwell's nagging concern over the sanctity of the school fund, along with several other competing interests, resulted in the proposal being tabled in the Senate and referred to the next legislative session.

Sensing that success was around the corner (since neither the 1859 or the 1860 proposal was voted down, merely tabled until the following legislative session), members of the Boston Society of Natural History decided to revive the discontinued practice of an annual address to the organization's members in the hope that it would evoke support for their idea of a Conservatory of Arts and Science in the Back Bay. Strategically, even presciently, they invited William Rogers to give the address. The meeting was held on May 11, 1860, to mark the thirtieth anniversary of the institution. Perhaps not surprisingly, it drew a large crowd, and true to form, William did not disappoint: he laid out the necessity of the consortium's proposal brilliantly, imploring attendees to support the plan when the Reservation Committee of the Associated Institutions presented it to the legislature during the upcoming legislative session.

To prepare for the legislative session, the governing body of the Associated Institutions trimmed membership of the Reservation Committee from eighteen members to five, choosing William as its chairman. What marked this committee was its keen interest in establishing a polytechnic institution for the industrial arts, which had been mentioned in the previous two proposals, but never fully explicated. This time, William intended to place that idea at the center of the proposal.[14]

Early in the summer of 1860, William and Emma retreated to Sunny Hill, Emma's father's summer residence in Lunenburg, Massachusetts. There, William worked on a variety of projects, but most of all on the proposal for the Associated Institutions. By late September, he had a draft prepared. Drawing on his "School of Arts" prospectus for the Franklin Institute in 1837, his "Plan for a Polytechnic School in Boston" for the Lowell Institute in 1846, his detailed knowledge of scientific institutions and societies in Europe, and his recent experience revising the consortium's proposal for the Massachusetts General Court, William prepared *Objects and Plan of an Institute of Technology including a Society of Arts, a Museum of Arts, and a School of Industrial Science, proposed to be established in Boston*.[15] The plan comprised three co-equal branches: a Society of Arts responsible for the collection, publication, and dissemination of research applicable to the industrial arts; a Museum of Industrial Art and Science aimed at collecting and preserving objects of prominent importance to the practical or industrial arts in order to better understand the relationship between science and industry; and a School of Industrial Science and Art—William's ever-evolving idea of a polytechnic institute—focused on systematic training in the industrial or applied sciences through lecture, discussion, demonstrations, and in-depth, hands-on experience in technologically-advanced laboratories.[16]

In a letter to Henry dated September 24, 1860, William described the plan and his presentation of it to the Associated Institutions' full assembly:

> My last visit to Boston was for the purpose of reading to a committee a pretty full outline of an Institute of Technology, to comprise a Society of Arts, an Industrial Museum, and a School of Industrial Science. My plan is very large, but is much liked, and I shall probably submit it, by request, to a meeting of leading persons in the course of a week or two, after which it will be printed in pamphlet form. The educational feature of the plan is what ought most to recommend it, and will, I think, be well appreciated.[17]

Several weeks later, not only did members of the consortium approve William's plan, but they also voted to have it printed and distributed to interested parties in the greater Boston area in order to "elicit the opinions and invite the cooperation of those best qualified to judge of the practical merits of the plan...."[18] Having honed his political instincts over the last legislative session, William requested that distribution of the pamphlet, which for the first time alluded to the proposed technical institution as the Massachusetts Institute of Technology, be delayed until after the forthcoming November election. William had learned that even legislators had priorities, and gaining reelection was one of the most important and time consuming, especially during a presidential election cycle under the threat of Southern secession.

William's plan shifted the emphasis from the public worth of a collection of museums from which the public could appreciate and learn to the emerging educational component that he had been wrestling with ever since he proposed a school for the industrial classes for the Franklin Institute over twenty years earlier. In the final paragraphs of the circulated plan William more than once argued that the classical liberal arts education of the nation's colleges, even with increased opportunities for scientific study, were both limited to the few—the sons of the upper classes—and taught in a manner that was inappropriate to training the scientific mind:

> It will be apparent that the education which we seek to provide, although eminently practical in its aims, has no affin-

ity with that instruction in mere empirical routine which has sometimes been vaunted as the proper education for the industrial classes. We believe, on the contrary, that the most truly practical education, even in an industrial point of view, is one founded on a thorough knowledge of scientific laws and principles, and which unites with habits of close observation and exact reasoning a large general cultivation.[19]

The *Objects and Plan* that William drafted and the Associated Institutions circulated called specifically for a School of Industrial Science and Art that would offer regular classes in various branches of the applied sciences and arts directed towards the needs and interests of the mechanic arts, engineering, architecture, manufacturing, and agriculture. And, as it turned out, highlighting the educational component of the proposal was consistent with the aims and goals of Massachusetts' new governor, John Albion Andrew, who included a section on "Practical Scientific Institutions" in his inaugural address on January 5, 1861, calling for legislators "to use all reasonable means to promote the spread of useful knowledge and especially to facilitate such practical scientific instruction as shall elevate while it invigorates the industrial arts."[20]

The governor's call was not an abstract wish: Gov. Andrew expressly identified the efforts of the Boston Society of Natural History and the Massachusetts Horticultural Society, i.e., the Associated Institutions, to establish a "collocation of institutions devoted to practical branches of art and science."[21] But he didn't stop there: Andrew went on to commend the material benefits, which the proposed Institute of Technology would add to previous proposals, enjoining the legislature to give the forthcoming—the consortium's third—proposal their utmost consideration. Knowingly or not, Andrew set in motion a rapid series of events that culminated in the desired goal—legislative action. To start with, the day after the governor's address members of the Associated Institutions encouraged interested parties to attend a meeting the following week to discuss the status of their proposal and to consider filing a separate petition for an act of incorporation for the technology institute.

A robust group of supporters met on a chilly Friday evening, January 11, 1861, in the Mercantile Library Association Building on Summer Street to discuss the petition. The first order of business was to elect William Rogers chairman for the evening and John Runkle, a graduate of Lawrence Scientific School and currently assistant to Benjamin Peirce, the group's secretary. Believing that the final petition would be received favorably by the legislature, William proposed that a plan of government for the Institute be prepared and put into operation as soon as the Institute was chartered. To this effect, the following "Act of Association of an Institute of Technology" was adopted and signed by thirty-seven supporters of the plan:

> We the subscribers, feeling a deep interest in promoting the Industrial Arts and Sciences as well as Practical Education, heartily approve the objects and plan of an Institute of Technology, embracing a Society of Arts, a Museum of Arts, and a School of Industrial Science, as set forth in the Report of the Committee; and we hereby associate ourselves for the purposes of endeavoring to organize and establish in the city of Boston such an Institution, under the title of "The Massachusetts Institute of Technology," whensoever we may be legally empowered and properly prepared to carry these objects into effect."[22]

From these actions, a Committee of Twenty was formed, to which William was immediately added as chairman. This body, which included some of Boston's leading educational, professional, and business men (including Runkle, who would become the Institute's second president, and Francis Storer, the Institute's first professor of general and industrial chemistry),[23] was to work with previous committees and organizations in order to further the petition of Back Bay lands and establish the Institute of Technology, which was now at the forefront of the Associated Institutions' proposal (although "Associated Institutions" no longer adequately described the consortium as several constituent members had dropped out). Within days, the Committee of Twenty forwarded two petitions to the state legislature: the first asked that the legislature grant a charter to the proposed technology institute; the second petition asked that "a continuous piece of land be set aside in the Back Bay for its use and for the Horticultural and Natural History societies."[24]

Again, sensing opposition from those who staunchly defended the sanctity of the school fund, members of the Committee of Twenty who addressed the legislature's Joint Standing Committee on Education over the course of the next few weeks emphasized the positive impact that the proposal would have on adjacent land values and, as such, would offset the loss of any parcels of land given to the consortium. But stiff opposition continued, both by George Boutwell and Joseph White, who replaced Boutwell as Secretary of the State Board of Education. But William and others chiseled away at their argument, not only attacking it on fiscal grounds, but on educational grounds as well, stressing the "educational capital" that would accrue to the Commonwealth over and above any fiscal rewards.[25] In early February, a member of the Reservation Committee informed the governor of a suggestion made by a member of the State Board of Education that a meeting with the Reservation Committee might help clear up the Board's apprehension about the plan in relation to the school fund. After some prodding, Gov. Andrew finally agreed, but with the proviso that only William speak in defense of the Associated Institutions' proposal. Andrew not only supported the proposal, but also had full confidence in William's persuasive oratory skills. In a letter dated March 9, 1861, inviting William to meet with members of the Board of Education, Andrew wrote:

> My dear Professor,—The Board of Education will meet next Wednesday morning. I hope you will come and advance the claims of the Natural History and Institute of Technology, but no one else should speak. Be thou the advocate. Take time enough. Cover the ground to suit yourself; several speakers would do harm; at least I fear they would. And you may say from me that I wish one complete argument, and that no other be made. Between ourselves I know *you* would have a powerful effect, left to yourself, and I fear some one else might come in and weaken it.[26]

William did just that, but it was still not enough. Opposition remained: the bone of contention—concern over diminished contributions to the school fund. More meetings were scheduled, letters written, impromptu discussions held, until finally, the House of Delegates passed the Associated Institutions' proposal on April 8, 1861, followed by the Senate the next day. "An Act to Incorporate the Massachusetts Institute of Technology, and to Grant Aid to Said Institute and to the Boston Society of Natural History" was signed into law by Gov. Andrew the day after that, appearing as Chapter 183 in the *Acts and Resolves of Massachusetts for the Year 1860-1861*.[27] Two days later, on April 12, 1861, confederate forces shelled Union-controlled Fort Sumter nestled in the protective waters of Charleston Bay, an act of aggression that signaled the start of the War Between the States.

But war or no war, the march toward reforming the nation's colleges continued. In the spring of 1862, one year into the war, founding members of the newly chartered Massachusetts Institute of Technology, convened their first meeting whereupon members—a rare mix of industrialists, scientists, and educational reformers—ratified the Institute's charter, adopted its bylaws, and chose its officers, electing William Barton Rogers the Institute's first president. As significant as these actions were, a more significant action that summer by the U.S. Congress overshadowed them—passage of the Land-Grant College Act of 1862, otherwise known as the Morrill Act, named after its principal sponsor Vermont congressman Justin Smith Morrill.

EVER SINCE HE WAS elected to represent Vermont in the U.S. House of Representatives in 1854, Justin Morrill, the son of a blacksmith, sought national sponsorship for a congressional act that would establish state land-grant universities with an emphasis on agricultural and mechanical science. A voracious reader and bibliophile, Morrill had very little formal education, dropping out to work in the trades. Through persistence and hard work, Morrill rose from tradesman to businessman to congressman. Sensing the need for change in the antebellum college classroom, specifically the need to introduce applied science into the traditional liberal arts curriculum in order to answer the call for more practical education for the industrial classes, the so-called "sons of toil,"[28] Morrill introduced the land-grant bill to facilitate the growing interest in science and science education.

Signed into law by President Lincoln on July 2, 1862, the Morrill Act directed the federal government to give each state that accepted its terms thirty thousand acres of federal land for each representative and senator in Congress. This enabled states "to sell, rent, or otherwise derive an income from the scrip to finance the establishment of a new institution or to support existing ones"[29] dedicated to the teaching of the agricultural and mechanic arts. According to Roger Geiger, the intent of the legislation was "to promote the liberal and practical education of the industrial classes,"[30] echoing Morrill's own words, "Being myself the son of a hard-handed blacksmith...I could not overlook mechanics in any measure intended to aid the industrial classes in the procurement of an education that might exalt their usefulness."[31]

Not every state jumped at the chance to cash in, however. One of the earliest to do so was Connecticut, when Yale's Sheffield Scientific School, under the persistent efforts of junior faculty member Daniel Coit Gilman (later president of Johns Hopkins University, the nation's first full-blown research institution), offered to function as Connecticut's land-grant recipient. With no other offer on the table, it became the de facto recipient, receiving state support from 1863 to 1893, until the grant was transferred to the newly chartered Storrs Agricultural School (known subsequently as the Storrs Agricultural College, the Connecticut Agricultural College, Connecticut State College, and finally, in 1939, the University of Connecticut).

In hindsight, passage of the Morrill Act, which was first proposed by Morrill in 1857, passed by Congress in 1859 but vetoed by President Buchanan, and finally passed and signed into law by President Lincoln in 1862, was a watershed moment in the history of science education, enabling states to establish agricultural and technical programs within their borders either at established or new institutions of higher learning. The "either at established or new institutions" would prove to be a bone of contention in many states, as it pitted traditional colleges against the need to organize new institutions that—in the eyes of many—could better address technical and scientific education. There is no better example of this tension than in Massachusetts, in the push-and-pull between Harvard University and the Massachusetts Institute of Technology.

As soon as the Morrill Act was signed into law, William Rogers wondered how the Institute of Technology could profit from state support under the act. At the same time that he was thinking about the how the Morrill Act could benefit his institution, Louis Agassiz of Harvard's Lawrence Scientific School was wondering the same thing for his pet project—the creation of a Museum of Comparative Zoology at Harvard. In a flash of brilliance, Agassiz coupled his request for state support under the Morrill Act with a $200,000 bequest left to Harvard by the estate of Benjamin Bussey for the purpose of establishing an undergraduate school of agriculture and horticulture. Although Bussey's wishes were made known in 1835, the year of his death, it was not until 1861 that the bequest was made available to Harvard, just in time for Agassiz to make his proposal to couple the land-grant funds to the Bussey bequest, a move largely intended to thwart William's competitive bid for the land-grant funds. In a private meeting with the governor, Agassiz suggested that Harvard acquire the Institute of Technology, thus combining the Lawrence Scientific School, the Massachusetts Institute of Technology, and the Bussey Institute of Agriculture into an amalgam of scientific and agricultural interests. Although a close associate of William Rogers, Gov. Andrew chose not to inform him of Agassiz's merger plans; instead, the governor asked William to write a report on the current status of the Institute and his future plans for it. In his request, Andrew benignly asked William what he thought about merging Harvard's scientific school with the Institute of Technology in order to obtain the Morrill Act appropriation.

William recoiled, stating in no uncertain terms that the Institute's independence was essential, especially from a predominantly classical liberal arts college such as Harvard, despite its forays into a professional school of science and technology, i.e., the Lawrence Scientific School. More than that, William felt strongly that the freedom to experiment with instructional approaches, the curriculum, and approaches to scientific research was essential to the lifeblood of the Institute of Technology and its future. But the real problem, as A. J. Angulo notes, was William's antipathy toward Agassiz, who had transformed the original applied science mission of the Lawrence Scientific School "into an extension of his research that had little to do with utility or application."[32] There were other points of divergence between the two brilliant scholars: their attitudes toward the professionalism of science, the nature of observation and experimentation in a science curriculum, the place of pure or theoretical science at advanced levels of instruction, and the nature and role of species classification and of museum work in general.[33] Given these factors, William could hardly envision a union between Harvard and his recently chartered Institute of Technology, even if it meant losing the Morrill Act appropriation.

Despite their differences, and William's plea to the governor not to consider such a merger, in his January 1863 address to the state legislature, Gov. Andrew sided with Agassiz, calling for a merger between the two institutions in order to create a "true university."[34] Although he made a convincing argument, Andrew failed to convince every state legislator, so no decision was made concerning the proposed merger in the final report. However, legislators did agree to accept the land-grant appropriation, to put the proceeds from land sales into a trust, and to distribute the interest from the trust accordingly: one-tenth of the proceeds to be used to buy land on which to build an agricultural college; and of the remaining interest, one-third to go to the Institute of Technology and two-thirds to go to developing the agricultural college, the latter being located in Amherst rather than Boston.

William's need for money, however, did not disappear. When the Institute was chartered in the spring of 1861, the state legislature added the stipulation that William and his supporters raise a guarantee fund of $100,000 within a year before the state would grant the Associated Institutions land in the Back Bay. With the outbreak of the Civil War that deadline, April 1862, was extended one year. Despite his success in securing state funds from the Morrill Act, Rogers still needed to fulfill his pledge to the legislature to raise $100,000 by the year-end extension, which was fast approaching. To this end, William made another round of pleas to his supporters, imploring recipients of his request to help meet the $100,000 fund-raising mark before the April 1863 deadline. Over the next few weeks, pledges of small donations trickled in: encouraging, but not enough to meet the required goal. Then on April 9, 1863, one day before the expiration of the extension date, a Boston-area physician and philanthropist stepped forward to fill the gap, pledging $60,000. William Barton Rogers now had in place a charter, a governing board, a site, an endowment, and pledges totaling $100,000—he just didn't have a building or students. Over the next eighteen months, William and a core group of supporters worked tirelessly to put into place the remaining pieces of this elaborate jigsaw puzzle.

CHAPTER 10 | *The New Education*

> *The Puritans thought they must have trained ministers for the Church and they supported Harvard College—when the American people are convinced that they require more competent chemists, engineers, artists, architects than they now have, they will somehow establish the institutions to train them.*[1]
> —Charles William Eliot

To help put the pieces of this large puzzle together, William Rogers tapped the expertise and experience of a number of Bostonians. He had to: there were just too many questions to answer and too much work to do. Organizationally, the Associated Institutions' proposal revolved around three components: the Society of Arts, the Museum of Technology, and the School of Industrial Science. To address the needs of these domains, thirteen committees were formed with the entire operation overseen by Rogers, who was elected president of the Institute at the governing body's first meeting in 1862.[2] Over the next three years, Rogers oversaw the work of the various committees that regularly met to iron out unresolved issues, such as: Where would classes be held before a building was built in the Back Bay? How would the Institute secure immediate operational funds? How would it secure an endowment to pay faculty salaries? How would the school be organized? Who should be hired? And what should they teach? These were no small matters, and they needed answers before the anticipated opening date at the beginning of February 1865.

Of everything that was demanded, the most important item was a strategic plan, a foundational document that would serve, at least for the moment, as the Institute's directional compass. As the Institute's first president, Rogers took up the pen and set out to write such a document, though it did not come easily. The result was *Scope and Plan of the School of Industrial Science of the Massachusetts Institute of Technology*,[3] which quickly became the blueprint of the Institute's educational mission, a mission that went beyond what Rogers had outlined for the state legislature in his *Objects and Plan* proposal. The document called for two departments, one emphasizing a general or popular course of study, and a second department focused on special or professional instruction. While the first department consisted of a series of introductory lectures on mathematics, physics and mechanics, chemistry and its applications, geology and mining, and botany and zoology, the second department was organized around five professional four-year courses intended "for students seeking a full course of scientific studies to fit them for the professions of mechanical, civil, or mining engineering, building and architecture, or practical chemistry."[4]

In a letter dated January 24, 1865, that circulated among several hundred of the Institute's supporters, Rogers wrote:

> The studies and exercises of the School are so organized as to provide a complete course of instruction and training, suited to the various practical professions of the Mechanician, the Civil Engineer, the Builder and Architect, the Mining Engineer, and the Practical Chemist; and, at the same time, to meet the more limited aims of such as desire to secure a scientific preparation for special industrial pursuits,—such as the Direction of Mills, Machine Shops, Railroads, Mines, Chemical Works, Glass, Pottery, and Paper Manufacturers, and of Dyeing, Print, and Gas Works; and for the practice of Navigation and Surveying, of Telegraphy, Photography, and Electrotyping, and the various other Arts having their foundation in the exact sciences.[5]

Reception of Rogers' plan was widespread and positive (it had already been approved by the Institute's governing body at a general meeting on May 30, 1864), and no one said it better than Rogers' closest associate and most vocal supporter, John Runkle:

I have analyzed [the plan] with the greatest care, carrying in imagination students through each of the courses from year to year, & I find it to my mind, perfect in all its parts. I am sure that in this country, (& I doubt if in any other, even in France, where the largest experience & study have been devoted to the subject), no institution has ever been based upon so comprehensive & perfect a plan; & I trust that the Government of the Inst. will not rest satisfied until they have secured every facility for its complete execution.[6]

Despite the unique organizational elements outlined in Rogers' *Scope and Plan*, what really distinguished the School of Industrial Science from other scientific schools—from the Military Academy at West Point, Rensselaer Polytechnic Institute in Troy, New York, the Sheffield Scientific School at Yale, Harvard's Lawrence Scientific School, and the recently endowed Abel Chandler Scientific School at Dartmouth[7]—were three essential ingredients: an unabashed emphasis on learning-by-doing in an experimental laboratory environment, the coupling of theory and practice, and the availability of a comprehensive program of study that allowed for degree specialization.[8]

The integration of these key ingredients into a structure that made sense did not occur overnight however. There were many structural elements to settle. In the end, however, Rogers and the Committee on Instruction settled on a four-year course of study that was comprised of two years of required courses and two years of specialized study in one of six areas. The six areas of study as listed in the first course catalogue were: mechanical engineering, civil and topographical engineering, geology and mining, practical chemistry, building and architecture, and general science and literature.[9] For a more complete summary of the Institute's goals, we turn to the annual course catalogue published at the start of the 1868-69 academic year:

> The Massachusetts Institute of Technology provides a four years' course of scientific and literary studies and practical exercises, embracing pure and applied mathematics, the phys-

> ical and natural sciences with their applications, drawing, the English language, mental and political science, French and German. The course is so selected and arranged as to offer a liberal and practical education in preparation for active pursuits, as well as a thorough training for the professions of the Civil and Mechanical Engineer, Chemist, Metallurgist, Engineer of Mines, Architect, and Teacher of Science.[10]

Two things stand out in this statement. First of all, by 1868, the term "School of Industrial Science," one of three components of the proposed Institute of Technology, was being used interchangeably with the name by which we know the institution today—Massachusetts Institute of Technology. Secondly, although evening classes were also offered, it was clear by the third full year of operation that the Institute's focus was on the specialized programs offered during the day to full-time students. At the heart of these programs was the primacy of the laboratory, but not for lecture-demonstrations by professors, rather for students to conduct their own experiments, first in a controlled manner as part of the first two years of required coursework, then for student-initiated projects conducted during the last two years of specialized study. Ultimately, this was what gave the Institute its distinctive character, setting it apart from other polytechnic and scientific schools, even its closest competitor, the Lawrence Scientific School.

To service the six departments, or divisions, four laboratories were planned in the following areas: physics and mechanics, general chemical analysis and manipulation, metallurgy and mining, and industrial chemistry. Along with laboratory work, students were required to sit for regular written examinations and complete a senior thesis on a topic of their choosing. Although laboratory instruction appears in the early planning stages, it was not expected that the four laboratories would be fully operational by the projected start date of February 1865. In fact, the Institute only opened thanks to an anonymous donor who agreed to pay rent on two rooms in the Mercantile Library Association Building on Summer Street while a suitable building was constructed on Boylston Street in the Back Bay. Although the two rooms in the Mercantile Building—one used for a general-purpose classroom, the other for a rudimentary chemistry lab—adequately accommodated the twenty-three students who completed the preliminary session, as the fall term approached, a steady stream of inquiries from parents indicated that the projected enrollment for the fall would be closer to seventy students. Anticipating a significant increase in enrollment, the Building Committee leased two floors of the Congregational Library Building on Chauncy Street. The question of space was also solved by an arrangement with the Lowell Institute, which agreed to hold some of the Institute's courses in the evening free and open to all.[11] Of course these were temporary arrangements until the Institute's new building was available on land adjacent to the Natural History Museum in the Back Bay.[12]

Money, buildings, curriculum, and equipment aside, the most important factor in founding the Institute of Technology was the selection of faculty. From the start, Rogers surrounded himself with some of the brightest minds in Boston's scientific and intellectual circles, with many supporters having deep connections to Harvard, especially to the Lawrence Scientific School. Heading the faculty list was William Rogers who, along with his administrative duties as president of the Institute of Technology and principal of the School of Industrial Science, would teach physics and geology. John Runkle, a graduate of the Lawrence Scientific School and a former student of Benjamin Peirce, would teach mathematics and analytical mechanics and serve as secretary of the Institute. In addition to Rogers and Runkle, the following faculty members were recruited for the preliminary session that began February 20, 1865, in the Mercantile Building: Harvard graduate William Watson (mechanical engineering and graphical works); Francis Storer, former assistant to Josiah Cooke (chemistry); W. T. Carlton, an instructor at the Lowell drawing school (drafting and freehand drawing); and Ferdinand Bôcher, Harvard instructor of modern languages (French language).[13]

It was a small but extremely robust group of individuals, who, in the earliest days of the Institute, shouldered the burden of this enormous task, a task that not only took the full measure of the faculty's time and energy (and just as often their money), but also demanded that they step outside of their role as "professor" and wear a variety of other hats, as the following citation suggestions:

> As a group they were in effect officers of admission, overseers of discipline, guardians of the building, schedulers of classes, proctors of the drawing rooms, supervisors of the drill, watchdogs for absences, monitors of academic performance—all this plus the lecturing, laboratory supervision, and examining duties related to their classes and the subjects which they had themselves designed. They dealt minutely with every detail of the catalogue and the curriculum, and revised and adapted their plans as experience dictated. As a group and in committees they were concerned with every aspect of the school, including additions to the instructional staff, and with every detail of its operations not the prerogative of the janitor.[14]

Additional faculty members were recruited by the fall of the following year: they included William Atkinson[15] (English language and literature), William Ware (architecture), Charles Eliot (analytical chemistry and metallurgy), John Henck (civil and topographical engineering), and James Hague[16] (mining engineering). Although each one of these men stood out in their own right (or would later in his career), one member of the group deserves closer attention.

CHARLES WILLIAM ELIOT WAS born into a family "with ancient, extensive, and important connections" both to the elite families of Boston, but also to Harvard College across the Charles River.[17] Hugh Hawkins makes these connections crystal clear in the opening paragraph of his biography of Eliot:

> For Charles William Eliot, entering Harvard College as a freshman in 1849 was hardly leaving home. His father, Samuel Atkins Eliot, was the college's treasurer and *ex officio* member of the Corporation; he had just published a history

of Harvard. Charles's grandfather, a highly successful import merchant and possibly the richest Bostonian of his day, had endowed the college's Eliot Professorship of Greek in 1814. Two uncles by marriage, George Ticknor and Andrews Norton, were former Harvard professors, internationally recognized scholars, and leaders in Boston-Cambridge intellectual circles. Harvard's new president, Jared Sparks, was an intimate family friend… To enter Harvard, then, was only to shift location slightly within the family.[18]

Like the scion of any wealthy Boston family, Eliot attended Boston Latin prior to matriculating at Harvard College. Although a rigorous intellectual experience, it was a social disaster—or could have been. Due to his appearance (Eliot was tall, lanky, extremely nearsighted, and carried a disfiguring birthmark on the right side of his face), Eliot endured the jeers and jabs of playground bullies throughout most of his Boston Latin years. However, indicative of his character, the protracted experience forced Eliot to take stock of himself, his family heritage, and his future, which he did, recalling that when he left Boston Latin, he was "reserved, industrious, independent and ambitious."[19]

Eliot entered Harvard College as a freshman in the fall of 1849 and, for the next four years, threw himself into his studies. Eliot had many memorable professors, but the most influential instructor was Josiah Cooke, Harvard's Erving Professor of Chemistry and Mineralogy. A graduate of Harvard with a year of study abroad, the enthusiastic and largely self-taught Cooke was entirely committed to both the advancement of scientific knowledge and the wellbeing of the college.

An excellent researcher, Cooke was also an exciting teacher, often supplementing daily recitations—the typical program of teaching in the academic college—with lecture-demonstrations. He also arranged excursions to nearby factories so his students could observe the principles of chemistry being applied in the world of commerce. Beginning in his sophomore year, Eliot studied with Cooke in his private laboratory in the basement of University Hall, and was soon assisting him and tending to the college's mineralogical cabinet. Recalling the influence that Cooke exerted over him, Eliot wrote: "I, for one, first learned what Chemistry was about, and what was the scientific method in observing and reasoning."[20]

Eliot learned more than what chemistry was about, however; he also learned about friendship and camaraderie. Along with assisting Cooke in the chemistry lab, Eliot often joined the young professor on trips throughout New England and New York in search of unique geological and mineralogical features. Another lab assistant—Francis Storer—often joined the two avid hikers. Over the years, Eliot and Storer would develop a deep and long-lasting friendship, not just as hiking companions, but also as colleagues plowing the same professional field—general and industrial chemistry.[21]

Like Eliot, who descended from notable family stock, Francis "Frank" Storer was the son of a Boston Brahmin. Dr. David Storer was professor of obstetrics and medical jurisprudence at Harvard for fourteen years and served as dean of the Medical School from 1855 to 1864 before retiring in 1868. The elder Storer's life was imbued with science: along with his professional interest in medicine, he was a distinguished naturalist, with special interests in reptiles and fish (he was also an avid collector of shells and coins). The Storer children grew up in a household in which science permeated the entire atmosphere, which explains why the younger Storer enrolled in the Lawrence Scientific School in 1850 at the age of eighteen, and in his second year, became one of Josiah Cooke's lab assistants.

After working together for several years in Cooke's chemistry lab, Eliot's and Storer's paths diverged. In 1853, Storer interrupted his studies to join the United States North Pacific Exploring Expedition as the expedition's resident chemist. Upon his return a year later, Storer resumed his studies at the Lawrence Scientific School, receiving his bachelor of science *summa cum laude* in 1855, whereupon he left for Europe and spent the next two years studying with some of the greatest chemists of the day, among them Robert Bunsen, Theodor Richter, and Émile Kopp. Upon his return, Storer established a private lab in Boston and worked as a consulting chemist, primarily for manufacturing, pharmaceutical, and commercial interests in the greater Boston area.[22]

Eliot, on the other hand, graduated from Harvard in 1853, second in a class of eighty-eight, but he didn't leave Cambridge: he stayed at Harvard after accepting a tutorship in mathematics under Benjamin Peirce. It was the fulfillment of a promise Eliot made to himself upon graduation: to blend "concern for a constructive role in society with concern for self-development."[23] Eliot had resolved to become a teacher, rather than a lawyer, a doctor, or a banker, expectations that his friends and family held out for him. As a tutor, Eliot quickly became known for being fair and effective, though somewhat remote and, as one student recalled, "cold as an icicle."[24] Remote or not, he was also an effective reformer, unafraid of trying new approaches, especially if he could make the subject matter more concrete. To this end, he borrowed surveying equipment and helped a group of students apply the trigonometry they were learning in class in a thorough survey of the college yard.

In 1858, Eliot was promoted to assistant professor of mathematics and chemistry, but his uniqueness on the faculty came not from his abilities as a teacher; rather they came from "his genius for organization."[25] And, by 1858, Eliot's abilities were on full display: he was registrar, secretary to the Corporation, secretary of the faculty, director of physical education, director of buildings and grounds, curator of mineralogy, librarian in charge of purchasing, alumni relations, and development director.[26] It is a testament not only to Eliot's industry and enthusiasm, but also to his native administrative skills.

Along with Cooke, Eliot's assistant professorship put him in contact with Eben Horsford, the Lawrence Scientific School's Rumford Professor of Applied Science. Eliot's relationship with Cooke and Horsford should have boded well for the aspiring chemistry professor. In fact, Eliot was counting on his connections—both at Harvard and through his family—in order to move forward in his career. Eliot had hoped that the next rung on his career ladder would be his appointment to the Rumford professorship, a position that became available in 1863, when Horsford resigned to devote more time to his private business interests. But the timing of Horsford's resignation did not favor Eliot. For one thing, Eliot's relationship with Cooke had been deteriorating for some time due to their differing views of theology and its role in scientific thinking, sparked by the 1859 publication of Darwin's *On the Origin of Species*. Whereas Cooke, like Louis Agassiz, Lawrence Scientific School's professor of comparative zoology, maintained a belief in "special creation," Eliot sided with Darwin, who wanted "to divorce scientific hypothesizing from theological considerations."[27] They also differed on approaches to teaching, with Eliot devoted to a new type of instruction based on hands-on laboratory experience and Cooke wedded to the lecture-demonstration approach. As tension between them mounted, Eliot—at his own expense—moved out of Boylston Hall, the site of Cooke's newly relocated laboratory, to a separate laboratory in University Hall. It was only when President Walker intervened that the two men reconciled their differences and acted amicably toward each other.

Horsford was a different problem. By 1861, after more than a decade at the helm of the Lawrence Scientific School, Horsford's results were not impressive: enrollment was stagnant, the school's debts were mounting, lab equipment went unaccounted for, faculty members were balkanized in their separate corners, and Horsford, overly involved with his own research, had begun to neglect his teaching duties. Under pressure from members of Harvard Corporation, Horsford handed over his administrative duties to Henry Eustis, head of the engineering department, but retained the position of Rumford Professor of Applied Science. The Corporation, on Eustis's and Cooke's recommendation, turned to Eliot to take over Horsford's teaching and laboratory duties. A year later, after Eustis went on leave for active duty in the Union army, Eliot took over as acting dean of the school. Eliot was now in a position not only to run the school, but also to suggest changes to its organization and curriculum. And he did so post haste: he ordered an inventory of all books and equipment at the school, putting in place strict faculty accountability for the school's resources; he scheduled regular faculty meetings every other week; he proposed more stringent entrance requirements; and he created a special committee to review the scientific school's curriculum. The committee, with Eliot firmly at the helm, proposed a four-year program consisting of a two-year general studies program followed by two years of degree specialization.[28]

At first, Agassiz showed little interest in Eliot's ideas, believing—like so many other reform initiatives—that they would come to naught. But Eliot was not only persistent; he was persuasive and gained the support of Joseph Lovering, Jeffries Wyman, and Asa Gray. With Peirce's backing, Agassiz voiced his concern about the program changes, at first mildly, then stridently. Agassiz's opposition was based on a combination of personal threat and utopian vision. On the one hand, Agassiz was extremely reluctant to give up any authority over his domain, especially at the suggestion of a junior faculty member, which Eliot was at the time. On the other hand, Agassiz belonged to a group of American scientists—the Lazzaroni[29]—that hoped to establish a "national university" that featured advanced lectures by the nation's most gifted and creative scientists and scholars (a university where personality trumped program). Having given up on the idea of founding a new institution, members of the Lazzaroni, headed by Alexander Bache of the United States Coast Survey, turned to Harvard, hoping to transform the historic college into their utopian vision. To Agassiz, Eliot's programmatic changes at the scientific school presented a structural barrier to that vision, and he let everyone know it, including Harvard's president Thomas Hill, who sided with Agassiz.

It was the confluence of Eliot's tenuous relationship with Cooke and Agassiz's triumph in squelching Eliot's reform proposals that spelled trouble for Eliot in 1863, when Eben Horsford resigned. Although it was reasonable to assume that Eliot, whose five-year appointment as professor of mathematics and chemistry was about to expire, would replace Horsford as the next Rumford professor (after all, Eliot had proven his worth to the institution many times over), Agassiz's antipathy toward Eliot (more than Cooke's tenuous relationship with Eliot) stood in the way—and it all came to a head during the search to replace Horsford.

Although he had his supporters within the Harvard community, Eliot was no match for Agassiz, who had supporters both in Cambridge and in the national scientific community. Agassiz also had the ear of President Hill, who often sided with the powerful and persuasive Agassiz (after all, it was Agassiz and Peirce who had lobbied hard for Hill's appointment in the first place). Although Eliot was considered for the Rumford professorship, ultimately it went to an outsider, Wolcott Gibbs, who was an extremely well qualified candidate currently teaching chemistry at the Free Academy of New York. Gibbs was the son of Laura Wolcott and Col. George Gibbs, a close associate of Benjamin Silliman, Sr. at Yale. Although Col. Gibbs was an enthusiastic mineralogist, whose "mineral cabinet" became the nucleus of Yale's great mineral collection and for whom the mineral *gibbsite* was named, it is Wolcott Gibbs' mother's family that stands out, with three generations of Wolcotts, starting with Laura's great-grandfather, Roger Wolcott, serving as governor of the State of Connecticut. Science on the one hand, politics on the other: in either case, it was a deeply intellectual and stimulating environment in which to grow up, one, like the Storer household, thoroughly dominated by "a taste for science."[30] Ultimately, that was what swayed Wolcott Gibbs' choice of profession.

After receiving his bachelor's degree from Columbia in New York, Gibbs went to Philadelphia to study with Robert Hare, who was a chemistry professor in the Medical School at the University of Pennsylvania. From Philadelphia, Gibbs returned to New York, where he received his medical degree from the College of Physicians and Surgeons. As many young men of his stature and family position did, Gibbs left for Europe to complete his advanced studies: in Germany under Rammelsberg, Rose, and Liebig, and in France under Laurent, Dumas, and Regnault. Upon his return to America, Gibbs delivered a course of lectures on chemistry at Newark College in Delaware and then, in 1849, accepted a professorship of chemistry at the Free Academy.

The appointment of Gibbs as Eben Horsford's successor, however, was not just a question of qualifications; it was equally a question of political influence. Agassiz and Peirce, founding members of the Lazzaroni, lobbied hard for the appointment of Gibbs, who was also a member of this elite group of American scientists who believed it their right to direct the course of American science. Two of the most prominent ways in which the Lazzaroni exerted their influence was through leadership in national scientific organizations and through appointments of candidates to influential professorships. In short, it was not Gibbs' stellar qualifications that sealed Eliot's fate; it was the fact that Gibbs was a member of the Lazzaroni and Eliot was not. So Eliot packed his bags and did what any wealthy young man from Boston would do: he set sail for Europe to continue his education.[31]

ALTHOUGH HE CAME FROM an established, well-heeled family, by 1858, the year he was promoted from assistant instructor to assistant professor, Charles Eliot's family's fortunes had changed drastically. A year earlier, due to some ill-timed investments, his father lost most of the family's wealth in the Panic of 1857. The younger Eliot's promotion to assistant professor of mathematics and chemistry in 1858 couldn't have come at a better time. But an assistant professor's salary was still a paltry sum and Eliot had a wife and child to support, which made the availability of the Rumford professorship even more appealing.[32] Fortuitously—perhaps even miraculously—at the same time Eliot received news that the Rumford position had gone to Gibbs, he received news that a small inheritance had come through from some investments on his mother's side of the family, enough to enable Eliot, his wife and child, and a nursemaid, to sail for Europe, which they did in September 1863.

For the next two years, Eliot explored the role of education in every aspect of European life: he attended public lectures, toured laboratories, visited polytechnic schools, met with leading chemists, and obtained copies of books, curriculum materials, and course listings. He did it with an eye on institutional organization and instruction, especially in the fields of science and technology. In France, Eliot made a particularly thorough tour of a number of institutions of note: the Conservatoire des Arts et Métiers, the l'École Centrale, the Sorbonne, the Jardin des Plantes, along with stops at various secondary schools with highly-acclaimed technical programs. In this crucible of travel and self-study, Eliot the teacher, the scientist, and the administrator were being forged into a larger, more complete entity, a professional educator with an overriding interest in the relationship between education and economic growth.

When asked by friends and colleagues at home how he spent his time, Eliot responded by saying that he was investigating schools that trained young men "for those arts and trades which require some knowledge of scientific principles and their applications."[33] Eliot's future was in education, in the advancement of science and technology, and he knew it—he just didn't know how or under what circumstances he would re-enter this domain once back in America. The first glimmer of life-after-study-abroad was an offer Eliot received from the Merrimack Company to supervise their large textile mill in Lowell, Massachusetts. It was a tempting offer: a considerable salary (twice that of a Harvard professor), a broad range of executive duties, and an elevation in status, especially among Boston's elite commercial class. But Eliot turned it down, confident that his future lay in the arena of education, not enterprise. It was not long after the Merrimack offer that Eliot received a second offer: this one came from a member of the Thursday Evening Club, William Rogers, who was several years into his own project—establishing an Institute of Technology in Boston's Back Bay.

After Rogers contacted Eliot about the possibility of joining the faculty of the fledgling institution, the two educators exchanged a series of letters, with Eliot asking questions on all matters of the Institute (its financial footing, faculty, curriculum, teaching methods, and physical space and equipment), and Rogers taking great care to answer each one of Eliot's concerns. Eliot also solicited the advice of some of his closest friends in Cambridge and Boston, including Frank Storer, who, by 1865, was among the new Institute of Technology's faculty. He also sought the advice of John Amory Lowell, a close family friend who was not only trustee of the Lowell Institute, but also served on the governing boards of both Harvard and the Institute of Technology. After weighing the responses he received from Rogers, and the enthusiastic letters he received from Storer, Lowell, and several other associates, Eliot accepted Rogers' offer to become Professor of Analytical Chemistry and Metallurgy at the Massachusetts Institute of Technology.

Eliot returned to Boston in the summer of 1865, ready to take up his responsibilities. From the beginning, with Storer his faithful companion, it was one large scramble to prepare teaching materials, outfit the chemistry laboratories, write curricula, recruit students, and, if this were not enough, prepare to move to the Institute's new Back Bay building, scheduled to open in the fall of 1866. But Eliot and Storer, both in their early thirties and full of energy and idealism, were a formidable duo:

> With Eliot the professor of analytical chemistry and metallurgy and Storer the professor of general and industrial chemistry, the two friends set to work. Together they planned the chemical laboratories in the new Institute building, where they accepted both degree candidates and a wide variety of special students. Their teaching was decidedly laboratory-centered, for both were convinced that "chemistry can only be learned tools in hand." The professor should not perform experiments before a large class. Every student should be in the laboratory, learning what it was to see with his own eyes, record his experience, and draw inferences from it.[34]

Their greatest work together, one in which they are still remembered, was the publication of a textbook of major importance: the landmark *Manual of Inorganic Chemistry*.[35] The 1867 publication, which quickly became a standard laboratory manual used at other colleges, articulated Eliot and Storer's innovative ideas about the role of hands-on learning in the training of young scientists. The source material for this book, of course, came from their many and varied experiences teaching in the chemistry laboratories both at Harvard and at the Institute of Technology. Organizationally, the lab manual included thirty-five chapters analyzing the chemical properties of elements and compounds, along with instructions for student experimentation and demonstrations of the facts and principles presented, with an appendix on "chemical manipulation" that explains the equipment and techniques to aid the student experimenter.[36] The first paragraph of the manual's preface states unequivocally the authors' intent:

> In preparing this manual, it has been the authors' object to facilitate the teaching of chemistry by the experimental and inductive method. The book will enable the careful student to acquaint himself with the main facts and principles of chemistry, through the attentive use of his own perceptive faculties, by a process not unlike that by which these facts and principles were first established. The authors believe that the study of a science of observation ought to develop and discipline the observing faculties, and that such a study fails of its true end if it becomes a mere exercise of the memory.[37]

The manual was an all-out attack on the status quo: on the lecture-demonstration method used in most science courses in nineteenth-century American colleges. The authors called for a new way of doing business in the college classroom, though by "new," they were not unaware of the fact that the experimental laboratory method they were promoting was "not unlike that by which [chemistry] facts and principles were first established."[38]

Despite his involvement at the Institute of Technology, Eliot could not help but be drawn back into Harvard's sphere of influence: in 1868, Eliot was elected to Harvard's Board of Overseers. On the one hand, it seemed an unlikely event: Eliot had left Harvard in 1863 both angered and humiliated by the appointment of Wolcott Gibbs as the fourth Rumford Professor of Applied Science. On the other hand, Eliot had his supporters: he came from admirable stock and he had, during the years he taught at Harvard, proven himself more than capable both as a teacher and an administrator. But none of that mattered until 1866, when the Massachusetts General Court voted in April to release private institutions from state control, which not only settled a long-standing dispute over the makeup of the Board of Overseers, but also made Harvard a distinctively private institution. The vote abolished public officials' *ex officiis* positions on the Board and, in doing so, transferred the power to elect Overseers to the alumni of the college "who voted in Cambridge and who were restricted in their choice of candidates to men who were residents of Massachusetts."[39]

Two things are germane to this discussion about the legislative action. First of all, after the vote, the Board created a committee to review the needs of the college, which determined that, faced with increasing competition from newly established polytechnic institutions, the college needed to adopt "all the best practical methods of education, whether they have been customary or not."[40] In other words, a thorough review of the "customary" classroom practices of recitation, disputation, and declamation were necessary, especially in light of the increasing presence of polytechnic and scientific schools, if the college were to thrive.

The second item of interest to this study is the election of Eliot to Harvard's Board of Overseers. As I mentioned earlier, the nomination, coming from a reinvigorated Board of Overseers, shouldn't be that surprising: after all, Eliot was a Harvard man (as was his father and most of his male relatives before him). Eliot had also demonstrated reform tendencies during his few short years at the emerging Institute of Technology in the Back Bay. In many ways, the gaunt, taciturn Eliot was a perfect addition to the Board given its recently adopted reform attitude. But it wasn't his presence on the Board that landed Eliot his next professional assignment. It was the publication of a two-part article in the February and March 1869 issues of *The Atlantic Monthly*, the nation's most prominent culture and opinion magazine.

In the article titled "The New Education," Eliot asks a simple question: "What can I do with my boy?"[41] It is a question, by 1869, that was on the minds of many educators about young men who demonstrated a decided propensity for practical work (what Holbrook would call the "mechanic arts" and Rogers the "useful arts"). By 1869, according to Eliot, the question had not been sufficiently answered by most traditional liberal arts colleges. After disparaging the current organization of the nation's colleges, many of them holdovers from the colonial era, Eliot suggested that what was needed was a new organization of schools—a "new education" or system thereof "based chiefly upon the pure and applied sciences, the living European languages, and mathematics, instead of upon Greek, Latin, and mathematics, as in the established college systems."[42]

Eliot then addressed the attempts at some colleges—Yale, Harvard, and Dartmouth—to create separate scientific schools. Although he recognized the good intentions of these creations, Eliot stressed their failures, which were not necessarily of their own making, but the result of an affiliation with a larger, more powerful, and established entity. After noting the dismal attendance and graduation rates of these so-called "foundlings," which he blamed in part on their extremely low admission standards and incomplete program offerings, Eliot wrote:

> The foundling has suffered by comparison with the children of the house. Even where there have been no jealousies about money or influence, and no jarrings about theological tendencies or religious temper, the faculty and students of the scientific schools have necessarily felt themselves in an inferior position to the college proper as regards property, numbers, and the confidence of the community. They have been in a defensive attitude. It is the story of the ugly duckling.[43]

Remembering his battles with Agassiz, Eliot was especially critical of Lawrence Scientific School, which in his mind was, and always had been, "what the Yale school also was at first, a group of independent professorships, each with its own treasury and its own methods of instruction."[44] His criticism was stinging and quite specific:

> This system, or, rather, lack of system, might do for really advanced students in science, for men in years and acquired

habits of study, in fact, the school has been of great service to a score or two of such men, but it is singularly ill adapted to the wants of the average American boy of eighteen. The range of study is inconceivably narrow; and it is quite possible for a young man to become a Bachelor of Science without a sound knowledge of any language, not even his own, and without any knowledge at all of philosophy, history, political science, or of any natural or physical science, except the single one to which he has devoted two or three years at the most.[45]

After a critique of separate scientific schools, Eliot moved on to discuss another approach to providing technical education at the post-secondary level, where two courses of study—one literary, the other scientific—ran side-by-side in parallel fashion, specifically citing Eliphalet Nott's initiatives at Union College. The plan was simple, though the execution was not always commensurate with the intention: a student chose one of two courses of study (either literary or scientific) and, at the end of his studies, received either a Bachelor of Arts or a Bachelor of Science, depending upon which course of study the student chose. The problem, as Eliot saw it, was that although in theory the two courses of study appear to be equal in nature, in reality, the institution, built upon the classical liberal arts model, inherently favored—in terms of faculty, resources, and student enrollments—the classical course and not the scientific course, thereby rendering the latter to an inferior, secondary status.

As he explored these two different approaches to providing scientific programs at traditionally established colleges, Eliot came to the conclusion that the two, at least as presently construed, must remain separate if, in fact, they were to become equal:

> The fact is, that the whole tone and spirit of a good college ought to be different in kind from that of a good polytechnic or scientific school. In the college, the desire for the broadest culture, for the best formation and information of the mind, the enthusiastic study of subjects for the love of

them without any ulterior objects, the love of learning and research for their own sake, should be the dominant ideas. In the polytechnic school should be found a mental training inferior to none in breadth and vigor, a thirst for knowledge, a genuine enthusiasm in scientific research, and a true love of nature; but underneath all these things is a temper or leading motive unlike that of a college.[46]

What, then, was Eliot's solution? According to Eliot, the two strands of education—the literary and the scientific—could coexist, but only if students were first offered a two-year general course of studies with a mix of literary and scientific course content that led students in their junior and senior years to an area of specialty that they freely chose. Unlike Francis Wayland at Brown University, who threw the barn doors wide open in his adoption of a full elective system, Eliot believed in a more controlled approach that led to a wider choice of subject matter in the last two years of study. This was Eliot's "ideal curriculum," a combination of an initial core of required courses followed by an elective system for upperclassmen, which would achieve the gold standard of education—mental discipline.

In other words, although he believed that the purpose of a liberal arts college was to teach men how to think, Eliot did not subscribe to the rigid view that the authors of the Yale Report held. And, shy of turning Harvard into a utilitarian technology institute, Eliot believed that the answer to the problem of how best to train men's minds while at the same time addressing the growing call and need for practical education was to replace the traditional liberal arts curriculum with a system that slowly introduced more electives, enabling students to take a mix of humanistic and scientific coursework. Eliot's argument was simple: if you could teach students to think with theology and Greek, why couldn't you teach them just as well with chemistry and German?[47] The question of subject matter went hand-in-hand with the question of pedagogy: when it came to teaching chemistry, geology, physics, or any other branch of science, Eliot believed that the best way to teach was in a laboratory environment, where students learned by trial-and-error and experimentation rather than by rote memorization. Eliot would echo this sentiment in his inaugural address in October 1869, stating that the best way to learn anything, but especially science, was "in a rational way, objects and instruments in hand,—not from books mostly, and through the memory chiefly, but by the seeing eye and the performing fingers."[48] It was a sentiment expressed two years earlier by Eliot and Storer in the preface of their laboratory manual in inorganic chemistry.

The motivation behind Eliot's call for reform, however, was not epistemological or pedagogical: it was pragmatic, based on the nation's need for more machinists, engineers, architects, chemists, and manufacturers. Eliot believed that what America needed, short of adopting the European university model in its entirety, was to create its own model of higher education—"the American university"—tailored to reflect the realities of America's unique and burgeoning needs.[49] In other words, Eliot's thesis aimed to "situate technical education in the current economic and social realities of American middle-class life."[50] As such, Eliot was not just an educational reformer; he was a social reformer, as his aim—indeed, his entire conviction—was to reform the system in order to create for the middle class the opportunities of the wealthy.

It was this conviction, along with his persuasive reasoning, his work at the Institute of Technology with Rogers and Storer, and his contacts at Harvard, that in March of 1869 brought Eliot the tantalizing offer to become Harvard's twenty-first president. But his ascendency to the prestigious position, the most prestigious position in American higher education, would not be easy: the offer split the Harvard community in half. The split is alluded to in the first sentence of Hugh Hawkins' chapter on Eliot's ascendency to the presidency: "Rarely in Harvard history had so much been at stake in the selection of a president."[51] What was at stake was the very survival of Harvard given its illustrious past but uncertain future. An uncertain future was precisely the prospect if Andrew Peabody, the acting president, was elected to replace the retiring Thomas Hill, who had stepped down at the end of September 1868. The paternalistic Peabody, the Plummer Professor of Christian Morals and Preacher to the University, would certainly guarantee a religious spirit at Harvard (a large part of Harvard's illustrious past)—but would he bring necessary reforms so desperately needed at the institution? That was the question and partly the reason for the split, which was further aggravated by the Board of Overseers' nomination of Eliot in March 1869, who by anyone's account did not fit the ministerial, paternal ideal: "there was suspicion that his theological liberalism verged on skepticism, since he was not among those scientists who dwelt on Nature's revelations of God's purpose."[52]

Even among Harvard's scientific community, Eliot's candidacy was viewed suspiciously: Did he really believe in a new American university, as he professed, or was he a college man? Then, of course, there was the question of his age (he was thirty-five years old) and his family's background (which was thoroughly aristocratic). In any case, Eliot proved to be a problematic choice, and yet he survived as he had done throughout his life when faced with seemingly insurmountable odds. After much back-and-forth between the Corporation and the Board of Overseers, on May 19, 1869, members of the Board of Overseers, by a vote of sixteen to eight, confirmed Eliot's appointment. Two months later, on July 1, 1869, Eliot submitted his letter of resignation to John Runkle, the Institute of Technology's acting president. But even in his brief four-year tenure as professor of analytic chemistry and metallurgy, Eliot left a legacy.

Aside from his regular duties with friend and colleague Frank Storer of designing, outfitting, and maintaining the chemistry laboratories, planning and teaching classes, and co-authoring an important and much-used laboratory manual in organic chemistry, Eliot put his administrative acumen to work, establishing a system of student bonding to guarantee against the loss of books and miscellaneous laboratory equipment, a process that he had hoped to establish at the Lawrence Scientific School when he was acting dean. Outside of the chemistry department, Eliot worked on several curriculum committees where he also had impact. In one committee, Eliot advocated for an initial two-year core curriculum of required courses that every student would take before declaring an area of specialization. Although the idea took root, it did not go unopposed, with his primary opponent John Henck, head of the civil engineering department, who argued for a narrow definition of the Institute as a professional school. In this regard, Rogers supported Eliot as he did Eliot and Storer's hands-on, learn-by-doing approach to chemistry. Finally, along with working closely with students in the chemistry and metallurgy laboratories, Eliot was also instrumental in advising Rogers on potential faculty hires and then taking an active role in mentoring junior faculty upon their arrival at the Institute. One of those junior faculty members was Edward Pickering.

In many ways, Eliot could have been thinking about Pickering when he wrote the opening paragraph to "The New Education" for *The Atlantic Monthly*:

> What can I do with my boy? I can afford, and am glad, to give him the best training to be had. I should be proud to have him turn out a preacher or a learned man; but I don't think he has the making of that in him. I want to give him a practical education; one that will prepare him, better than I was prepared, to follow my business or an other active calling. The classical schools and the colleges do not offer what I want. Where can I put him?[53]

Eliot would help Pickering answer that question several times in his young professional life. We've already seen Eliot intervene twice in Pickering's life. The first time occurred when Pickering was a student at Boston Latin: it was Eliot who suggested to headmaster Francis Gardner that Pickering would thrive best at the Lawrence Scientific School rather than Harvard College. The second time occurred when Pickering was at the scientific school: at the start of Pickering's second year of studies, Eliot recommended that Pickering change his major from chemistry to physics. It would not be the last time that Edward Pickering's self-appointed mentor would intervene in the young scientist's professional life.

CHAPTER 11 | *With His Own Hands*

It is well known that chemistry can be taught far better by a laboratory in which the student performs the various experiments, than by any system of lectures. Now, although for many years physicists have been in the habit of instructing their special students and assistants in this way, yet it is only recently that the same plan has been tried with large classes in physics.[1]
—Edward Charles Pickering

In February of 1867, Edward Pickering took up his responsibilities as assistant instructor of physics at the Massachusetts Institute of Technology in Boston's Back Bay. Pickering, who graduated from Lawrence Scientific School *summa cum laude* in 1865, was an assistant professor of mathematics at his alma mater when, on the advice of Charles Eliot, Rogers plucked the young instructor from Harvard's scientific school. The Institute of Technology had just moved from the Mercantile Library Association Building on Summer Street to its new home in the Back Bay next to the Natural History Museum on Boylston Street. If Edward Pickering was excited about taking up his new responsibilities, even as an assistant instructor under Rogers, it was because of the prospect of teaching physics in a laboratory setting.

Originally, the Institute's laboratories—four in all—were planned for a separate building, but financial constraints prohibited that accommodation, one of many problems facing construction of the Boylston Street building. Just the selection of an architect proved problematic, with the award going to the father-and-son team of Jonathan and William Preston rather than the firm of Gridley Bryant and Arthur Gilman, the more well-known of the two Boston firms. Of course, it didn't hurt that the elder Preston was chairman of the City of Boston's Commission on the Back Bay Lands. More than that, since the younger Preston was the lead architect for the Natural History Museum next door, it made sense to choose the father-and-son team to design the building for the Institute of Technology.

Despite the difficulties in selecting an architect, the shifting membership of the Building Committee, troubling cost overruns, and an uncertain completion date, the Rogers Building—as it would be officially named in 1883, one year after Rogers' death—opened in the fall of 1866, a year behind schedule and not quite complete. The massive neoclassical stone-and-brick building was a perfect match for the Natural History Museum next door, at least externally. Internally, it was a barren hulk of a building whose unfinished, raw interior only amplified the difference between it and the Natural History Museum next door. On paper, however, the layout of each floor signified a flurry of activities envisioned for the building. To see what was to come, let's fast-forward several years and take a tour of the building.[2]

Ascending the front steps of the building that face Boylston Street, visitors enter the building greeted by an enormous entry hall: on the left is the president's office, on the right the secretary's office. Proceeding around the entrance hall and its central stairway, visitors pass the Society of Arts lecture room, the mining and geological lecture room, and three rooms for the physics department: two laboratories and a lecture room. A quick trip to the basement introduces visitors to several chemistry and mineralogical laboratories and an adjacent lecture room. There is also a workshop for fabricating tools and small instruments, various storage and supply rooms, and, of course, the building's boiler room.

Ascending the stairway to the second floor, visitors are greeted by the entrance to the Great Assembly Hall, named Huntington Hall after benefactor Ralph Huntington. The cavernous hall takes up the full back half of the building and is a story-and-a-half tall. Walking in clockwise fashion around the main hallway in the middle of the building, visitors pass the astronomy lecture room, the English and modern language lecture rooms, the civil engineering lecture room, and the mathematics lecture room. Ascending the stairway to the next floor, visitors emerge into the "half story" wedged between the second and third floors created by the volume of the Great Assembly Hall. Standing at the top of the stairs, looking either left or right, visitors encounter two study areas for full-time professors. Walking around the main hallway in the middle of the building, visitors pass the natural history lecture room and museum, the architectural library and study room, a plastic modeling room, and two adjacent rooms of the architectural museum. The third floor is taken up by various freehand drawing rooms: one each for first-, second-, third-, and fourth-year students. There is also an architectural drawing room, a modeling room, a mechanical engineering lecture room, and a descriptive geometry lecture room. Finally, the fourth story contains half a dozen professors' study rooms, a study room for their assistants, another freehand drawing room, and a photographic laboratory.

Despite a statement in the first annual catalogue declaring that laboratory classes would be conducted in a small-class setting, there was very limited space for any of the laboratories and few, if any, equipment when the building opened in the fall of 1866. In the case of the physics department, the catalogue described offerings in a variety of mechanical and physical processes and materials, but with the qualifier that they would be included in laboratory instruction "once the proposed Laboratory of Physics and Mechanics had been established."[3] A tour of the building during its first year would have confirmed this, as there was no laboratory space for physics at all. In fact, very little effort had been made to establish a physics laboratory by the building's opening. The reason was twofold: not only was President Rogers overwhelmed by teaching and administrative duties, but also—and perhaps more importantly—the institution had fallen grossly into debt. More than likely, the two conditions combined prompted Rogers to hire Edward Pickering to free him from his teaching responsibilities so he could concentrate on the institution's pressing administrative and financial needs.

Upon taking up his appointment in February of 1867, Pickering immediately began substituting as a lecturer for Rogers, and within a few months had assumed most of Rogers' teaching duties, which provided the overworked and ailing president a much needed reprieve. At the same time, Pickering began to plan for the proposed physics laboratory. To this end, he commandeered a small room at the back of the architecture museum on the half-story floor between the second and third full floors, using it primarily with upperclassmen (in much the same way that faculty at other colleges did, inviting advanced or special students to work closely with them in their private laboratory).

By the winter of 1868, there was still no mention of an adequately outfitted physics laboratory. Nonetheless, it was a significant year. First of all, Rogers took a leave of absence due to declining health. Since his student days at the College of William and Mary, Rogers suffered from a weak constitution, often having to take a break from his unusually heavy workload. His replacement was John Runkle, the Institute's secretary and Rogers' closest confidant. Runkle was named acting president until Rogers was well enough to return. It was a natural choice since Runkle shared Rogers' vision for the fledgling Institute of Technology.[4] Along with Runkle's appointment, Edward Pickering was promoted from his position as assistant professor of physics (to which he had been rather quickly upgraded upon his arrival) to the newly established Thayer Professor of Physics, a title he would assume until he left the Institute eight years later.[5]

Inspired by his promotion, Pickering went to work outlining his intentions for the physics department and its laboratory in a document titled "Plan of the Physical Laboratory," which he submitted to Runkle and members of the governing board the following year. Although Pickering prepared the document, Rogers, even in absentia, provided the inspiration: in no uncertain terms, Rogers stated that physics should be taught in much the same way as chemistry was taught—in a fully-equipped laboratory using a hands-on, investigative approach.

After receiving Pickering's plan for the physics department, Runkle wrote to Rogers in the spring of 1869 to inform him of the young professor's progress:

> Pickering has drawn, in quite full detail, a plan for the physical laboratory, which I will send you before long.... Pickering is very anxious to be ready by October next to instruct the third year's class by laboratory work; and if an experience of one year shall be favorable, as I feel it must be, we can then gradually enlarge our facilities and take in the lower classes. I am convinced that in time we shall revolutionize the instruction in physics just as has been done in chemistry.[6]

When classes resumed in the fall of 1869, the physics laboratory, as the original floor plan indicated, occupied three rooms on the first floor of the Rogers Building (two rooms for laboratories and an accompanying lecture room). Pickering named the facility the Rogers Laboratory of Physics in honor of the Institute of Technology's founder. Anticipating a surge in student enrollment, Pickering worked out a pedagogical approach that seemed plausible given the size of the laboratory, the availability of equipment, and the number of students anticipated in any one class.

Essentially, Pickering had a choice between two methods of conducting large laboratory classes. One approach was to let all students perform the same experiment simultaneously, each student being supplied with all the apparatus necessary for the experiment. The second method was to let each student perform a different experiment, so that at any one time, there were a number of different experiments in progress as there were students. Science historian Florian Cajori weighs the pros and cons of these methods in his historical survey of laboratory instruction:

> The first method has the great advantage of permitting teachers to discuss, once for all, the theory of the experiment with the classes as a whole, instead of repeating it with each student individually. Moreover, it is easier to superintend a

large class when all are working at the same thing than when each is performing a separate task. The great disadvantage of this mode or procedure is that few institutions, if any, have the resources to furnish each student of a large class with the same instrument of precision. Several hundred instruments of the same kind might be needed for each experiment.[7]

Cajori notes an additional disadvantage of this approach: in order to afford the cost of a bulk order of laboratory equipment, the quality of the instruments often suffers, which in turn negatively impacts the degree of accuracy desired. Cajori then explores the pros and cons of the second method of large class instruction:

> The strong point of the second method is that it necessitates no duplication or multiplication of apparatus, thus making it easier to equip the laboratory with instruments of higher quality. Each student is at a different task. The members of the class rotate from one experiment to another on successive days. There is less opportunity for students to compare results, each pupil being thrown more upon his own resources. It is an individual method, calling for a great deal of "elbow instruction." A teacher cannot at one time take care of as many pupils by this method as by the first.[8]

There is a middle ground, however: a modified or hybrid approach that entails dividing the class into small groups and having each group conduct the same experiment at their assigned table, but with each table outfitted for a different experiment. If class time permits, student groups rotate through several experiments before the end of class, with time to discuss each table's results with the large group. An unintended consequence of this approach is that it engenders collaboration among small group members while, at the same time, fostering healthy competition between groups. Additionally, with less experiments to complete, the instructor can discuss the procedures and results of the various experiments with greater frequency either in small groups or with the entire class.

From the contents of an article Pickering published in the scientific journal *Nature*, it appears that Pickering preferred that each student work alone on a particular experiment, rotating among several experiments if time permitted. In the *Nature* article, Pickering offers the fullest account of his intentions for laboratory instruction, a process that he would constantly refine—and other institutions would eventually adopt:

> Two large rooms (one nearly 100 feet in length) are fitted up with tables, supplied with gas and water, somewhat like a chemical laboratory. On each is placed the apparatus prepared for a single experiment, which always remains in this place, thus avoiding the danger of breaking it in moving. A full written description is also given of each experiment, pointing out the proper precautions to avoid error or breakage. Near the door is an indicator or board containing the names of the experiments, and opposite each is placed a card bearing the name of the student. When the class enters the laboratory, they go to the indicator, and each member notices what experiment is opposite his name; he then goes to the proper table, reads the description, and performs it. He next reports his results to the instructor in charge, and if they are correct, his card is moved to some unoccupied place, and he proceeds as before. Care is taken that the number of experiments shall exceed that of students, and there is therefore no delay. The instructor in the mean time is enabled to pass from student to student and to see that no errors are committed.[9]

Whereas the above describes Pickering's pedagogical approach, in a report to President Runkle in the spring of 1870, Pickering reflects on the broad purposes of laboratory instruction in general:

> Its first object is to enable the regular students of the Institute, after attending a course of lectures on Physics, to verify its laws and measure its constants, also to learn to use the more important pieces of apparatus. Secondly, to instruct special students in the use of particular instruments, or branches of physics, as the spectroscope, microscope, pho-

tometry, electrical measurements, etc. Thirdly, to prepare teachers of this science. And fourthly, to afford facilities to physicists to carry on investigations at the Institute.[10]

In the same report, as he extolled the virtues of laboratory instruction, Pickering noted how it was "scarcely necessary" to point out the advantages of such an approach since it was currently the tendency of most, if not all, technical education in the country.[11] And it was: due in large part to William Rogers' vision and to Edward Pickering's persistence and dedication. That dedication enabled him—perhaps even drove him—not only to expand laboratory instruction to all levels of students, but also to expand the scope of laboratory work in general, as Julius Stratton and Loretta Mannix observe:

> He decided also to extend the scope of the laboratory work to assigned experimental investigators, followed by an opportunity for those who wished to pursue more advanced work to carry out original research and to publish their results. He had encouraged this from the start, and hoped that the laboratory would serve not only for teaching purposes, but also as a research facility.[12]

Pickering expounded on this crucial idea several years later in another end-of-year report to President Runkle:

> The great object to which my work for these last ten years has been directed, has been original investigation. This is the real goal to which the attention of a student in the Physical Laboratory is directed. The value of the course in physical manipulation is largely that it teaches a student to think for himself, and prepares him with methods by which he may solve any problem for himself experimentally. I have endeavored to impress on all our students in physics the principle that original investigation should be the great aim of every scientific man.[13]

It is a compelling vision, one that would influence the development of many scientific programs—undergraduate and graduate—at colleges

WITH HIS OWN HANDS 219

and universities across the nation, perhaps most notably Baltimore's Johns Hopkins University founded in 1876.

READING THE DESCRIPTION OF the contents and processes of the physical laboratory found in several of Edward Pickering's writings, I became interested in finding a visual legacy of the physics laboratory before Pickering left the institution in early 1877. Perhaps there was a drawing or a photograph of the laboratory when it was either at the back of the architectural museum on the building's half-story or on the first floor of the building as originally proposed. An Internet search quickly revealed a number of black-and-white photographs of the physical laboratories in the Walker Building that opened next to the Rogers Building in 1883 several years after Pickering had left the institution. I did find one image—a lithograph—of a laboratory in the Rogers Building on a webpage hosted by the Massachusetts Institute of Technology's Department of Archives & Special Collections. The lithograph was titled "Rogers Laboratory of Physics."[14] The accompanying caption stated that it was the physics laboratory in the Rogers Building ready for use in the fall of 1869, which would have been several months after the governing board approved Edward Pickering's "Plan of the Physical Laboratory." I should trust the veracity of the caption, but since several of the photographs taken in the Walker Building were incorrectly labeled, I thought I'd do a little detective work first.

The lithograph depicts one corner of a laboratory containing several tables, each table populated with some sort of apparatus or equipment. While sunlight floods the room from two tall, arched windows, three men and a woman stand off to the side. The windows are the key. If the laboratory was the one that Pickering had set up in the back of the architecture museum on the half-story between the second and third floors of the Rogers Building, the windows would not have been tall and arched: they would have been small and either oval or rectangular in shape. Tall, arched windows appear only on the first floor of the Rogers

Building, which places the laboratory in the lithograph on the first floor as per the original floor plan. However, since there were two rooms for the physics laboratory—as well as a lecture room—which one of the rooms is depicted in the lithograph?

To answer this question, let's turn to page fifty-six of Samuel Prescott's book, *When M.I.T. was "Boston Tech," 1861-1916,* that depicts a floor plan of the first floor of the Rogers Building. The key at the bottom of the page indicates that there are two adjacent laboratories for physics, one large and one small. Whereas the large laboratory spans the entire back of the building and encompasses three outer walls and their respective windows, the smaller laboratory, about a third of the size, encompasses only one outer wall, the rest being interior walls.

From this information, we can deduce that we are looking at one end of the large physics laboratory at the back of the Rogers Building, given that two out of three walls contain tall, arched windows. Furthermore, from the placement of the windows, and the sunlight flooding in, it appears that we are looking at the back northwest corner of the building (and only about a third of the laboratory's full size). Now let's look closer at what is contained in the laboratory. In the middle of the room, surrounded by eight-foot-tall glass storage cases, stand several tables laden with equipment. On one table, there's what appears to be a large electric fan. But it's not: it's the large glass disk of an electrostatic plate machine, a device invented in the eighteenth century for generating static electricity useful for any number of experiments (Pickering was enthralled with new developments in electricity; after all, it was the age of Edison, Tesla, and Bell). A bank of batteries sits on the floor in front of the electrostatic plate machine. Nearby, a "phonautograph," an early sound-recording machine used for the visual study of sound waves, takes up the surface of a small table.[15] To the left of the phonautograph is a column with equipment used for taking hydrostatic measurements. Finally, as I mentioned earlier, the lithograph depicts four individuals standing off to one side of the room: three of them, two men and a woman, listen to a third man who appears to be answering a question.

But what might the three onlookers have asked him? Perhaps they want to know what goes on in the rest of the laboratory, which the viewer does not see. We have already had a glimpse of what goes on in the laboratory as Pickering wrote about it, both in his report to President Runkle in 1870 and in the article published in the journal *Nature* the following year.

By the early 1870s, the Institute of Technology's physics laboratory, under Edward Pickering's careful direction, would have been humming with activity. Equipment was acquired, additional courses planned, and an approach to laboratory instruction was beginning to take shape, first sketched out in Pickering's "Plan of the Physical Laboratory," followed by his detailed description of laboratory procedures in the scientific journal *Nature*, and, finally, fully explicated in *Elements of Physical Manipulation*, a laboratory manual published by Pickering in two volumes in 1873 and 1876 respectively.[16] The inspiration for the work, which was not unlike Eliot and Storer's *Manual of Inorganic Chemistry* published several years earlier, was made obvious on the dedication page: "To Prof. Wm. B. Rogers, the first to propose a physical laboratory, this work is most affectionately inscribed by his sincere friend and pupil. Edward C. Pickering"[17] The manual begins by recognizing the proliferation of laboratory teaching and by acknowledging the author's experience working with students in a laboratory setting:

> The rapid spread of the Laboratory System of teaching Physics, both in this country and abroad, seems to render imperative the demand for a special text-book, to be used by the student. To meet this want the present work has been prepared, based on the experience gained in the Massachusetts Institute of Technology during the past four years.[18]

Organizationally, the book begins with a discussion of general methods of physical investigation that includes sections on the analytical method, the graphical method, and commentaries on different ways of taking physical measurements (i.e., time, weight, length, volume, angles, etc.). This is followed by a discussion of basic experiments and the care and calibration of equipment (i.e., insertion of crosshairs, suspension by silk fibers, testing thermometers, cleansing and calibration of mercury, the Hook gauge, and the estimation of tenths of a second, etc.). The remainder of the manual presents experiments by type of physical states, i.e., solid, liquid, gas, sound, and light (the latter two qualities of particular interest to Pickering). These were to be conducted in the laboratory by each student under the watchful eye of the instructor, but first, the student needed to understand two aspects of each experiment:

> Each experiment is divided into two parts; the first called *Apparatus*, giving a description of the instruments required, and designed to aid the instructor in preparing the laboratory for the class. The student should read this over, and with it the second part, entitled *Experiment*, which explains in detail what he is to do.[19]

But why teach this way in the first place? Pickering addressed this question in the next paragraph: "Perhaps the greatest advantage to be derived from a course of physical manipulation, is the means it affords of teaching a student to think for himself."[20] "To think for himself": herein lies the rationale for laboratory instruction (or, as Pickering calls it, "physical manipulation"), which coincidentally is not unlike what traditional educators had argued for in their staunch defense of the classical liberal arts curriculum decades earlier. They, too, wanted a curriculum—the classical liberal arts curriculum—that would teach students to think for themselves, or at least afford them a modicum of "mental discipline." Rather than achieve mental discipline through a set of daily rote exercises, Pickering—and Rogers, and Runkle, and Eliot, and others at the Institute of Technology—believed that to engage the mind, one must engage the hands. Pickering expresses this idea in the following lines:

> This should be encouraged by allowing [the student] to carry out any ideas that may occur to him, and so far as possible devise and construct, with his own hands, the apparatus needed.... To aid this work, a room adjoining the laboratory should be fitted up with a lathe and tools for

working in metals and wood, as most excellent results may sometimes be attained at very small expense, by apparatus thus constructed by students.[21]

This paragraph is useful not only because it helps us understand the importance of the students' engagement in the entire process, from fashioning their own instruments when necessary to the execution of an experiment, but also because it helps us understand the layout of the physics department. As you recall, the floor plan for the first floor showed a large laboratory at the back of the building, an adjacent smaller laboratory, and a lecture room. From the paragraph above, more than likely, students and faculty used the smaller laboratory as a workshop either to recalibrate instruments or to fashion new ones from scratch, reserving the larger room for student and faculty experiments conducted on tables outfitted with various equipment.

In *Elements of Physical Manipulation*, Pickering offers a detailed account of the methods and materials of laboratory instruction:

> The method of conducting a Physical Laboratory, for which this book is especially designed, and which has been in daily use with entire success at the Institute, is as follows. Each experiment is assigned to a table, on which the necessary apparatus is kept, and where it is always used. A board called an indicator is hung on the wall of the room, and carries two sets of cards opposite each other, one bearing the names of the experiments, the other those of the students. When the class enters the laboratory, each member goes to the indicator, sees what experiment is assigned to him, then to the proper table where he finds the instruments required, and by the aid of the book performs the experiment.... As soon as the experiment is completed, he reports the results to the instructor, who furnishes him with a piece of paper divided into squares if a curve is to be constructed, or with a blank to be filled out, when single measurements only have been taken. In either case a blank form is supplied, as a copy. New work is then assigned to him by merely moving his card opposite any unoccupied experiment. By follow-

ing this plan an instructor can readily superintend classes of about twenty at a time, and is free to pass continually from one to another, answering questions and seeing that no mistakes are made."[22]

But was Edward Pickering the first to teach physics in a laboratory environment by the investigative method? According to Rogers, he was, and he praised him as such in a letter to the Institute of Technology's governing body in 1870:

> It gives me pleasure to refer to the ability with which Prof. Pickering has carried out my views of a Laboratory of Physical Instruction. This is, I believe, the *first* laboratory of the kind ever established, and, as it furnishes practical training not hitherto attempted in any systematic way, will give the students of the Institute peculiar advantages in their studies, and in future researches in this very important branch of Science.[23]

In other written communication, Rogers eulogizes the work of the Institute, especially in the groundbreaking work of its laboratories: "Our institute may thus, I think, in [laboratory instruction] as well as in other features of its organization, claim the credit of having made an advance in practical scientific education."[24]

But was the Massachusetts Institute of Technology really the first to lay claim to such an innovative approach? According to Julius Stratton and Loretta Mannix, that honor goes to Rensselaer Polytechnic Institute under Amos Eaton, where as early as 1824, the school emphasized the laboratory method of instruction. At Rensselaer, professors gave lectures and conducted examinations, "but each student carried out experiments of his own design followed by lectures to his classmates explaining procedures and discussing results."[25]

Amos Eaton might have been the first to teach in an investigative laboratory manner, but it was Edward Pickering who fully fleshed out the approach involving students, faculty, and outside researchers in the proceedings of the laboratory with an eye on publishing and disseminating the obtained results. In other words, under Pickering, the physics laboratory at the Massachusetts Institute of Technology was not just a training ground for students, rather, it was an efficiently-run scientific community—albeit small—that nurtured a true collegial atmosphere, as Pickering himself observed in his report of progress made by the physics department for the *President's Report* for the year ending September 30, 1876:

> A good test of the character of the instruction, is the relation of teacher to pupil. My own relation, which have always been of the pleasantest character, I ascribe to the interest of the classes in their work. They remain beyond the hour, encroach upon their dinner-hour, and frequently ask, and are allowed, to work at other times which they might devote to amusement. With this condition of things disorder is almost unknown, and they are treated, and I believe regard themselves, as friends and guests, rather than as pupils.[26]

Edward Pickering began his tenure at the Massachusetts Institute of Technology on February 1, 1867; he ended it ten years later to the day on February 1, 1877. On that day, Edward Charles Pickering began the next—and, as it would turn out, final—leg of his long and distinguished professional career when he assumed the directorship of the Harvard Observatory. His work at the Massachusetts Institute of Technology garnered much attention, but no more so than from Pickering's long-time mentor and friend Charles Eliot who, as Harvard's twenty-first president, lured Pickering away from the Institute of Technology to direct Harvard's astronomical observatory, at the time one of the most important observatories in the world.[27]

BEFORE WE BRING THIS study to a close, there's one final image to mull over: it's a photograph of Edward Pickering taken a few years after he joined the faculty of the Massachusetts Institute of Technology. I've seen the photograph in a few books and on several websites describing the early years of the Institute. The photograph I'm looking at appears on the same webpage mentioned above titled "Rogers Laboratory of Physics." In the lower left-hand corner of the webpage is a black-and-white photograph of Pickering dressed in the gentleman's attire of the day: white collared shirt, floppy bowtie, buttoned waistcoat, and open frock coat. It's a curious headshot: the wan, lightly bearded Pickering stares blankly into the distance, lost in some unresolved piece of business. Is he thinking about a complex physics problem, the research of one of his students, or perhaps some much-needed laboratory equipment? It's hard to say. What is evident in this portrait, and in another photograph taken about the same time, is Pickering's delicate, almost ethereal, presence.

In a tribute to Edward Pickering after his death in 1919, Annie Cannon, an astronomer at the Harvard Observatory under Pickering, offers a glimpse of Pickering as an impressionable youth:

> Interested in the stars as a boy, when about twelve years old, he constructed his first telescope from an old pair of spectacle glasses. Dissatisfied with the result, however, he searched the shops on the Hill and found a small second hand lens, from which, with the aid of a piece of stovepipe, he made a telescope which showed Jupiter's satellites.[28]

How many boys—and girls, for that matter—were like Edward Pickering as a youth, interested in the world about them: curious, inventive, and persistent. Once again, it makes me think about the first line of Charles Eliot's article "The New Education" in which he asks the simple question, "What can I do with my boy?" He asks that question because he sees in his son—in this case, his mentee Edward Charles Pickering—something other than the usual professions of doctor, lawyer, minister, or teacher. He sees that he is practical, curious, good with his hands, and inventive. Is manual labor in his future—perhaps? But so maybe is engineering, or architecture, or manufacturing, or any number of other professions in the practical or useful arts. Eliot's "son," moreover, could have been any number of individuals—Sylvanus Thayer, Amos Eaton, Timothy Claxton, Benjamin Greene, Josiah Holbrook, Frank Storer, William Rogers—individuals with a talent and curiosity for the world and a demonstrated passion for new developments in science and technology.

That is precisely what the nineteenth century presented to many Americans: a world of endless possibility fueled by an explosion of technological innovation and commercial expansion. Edward Pickering understood this and grasped every opportunity that came his way. Like his father before him, Pickering climbed the ladder of success through a lifetime of hard work and dedication. What enabled him to climb higher than any other member of the Pickering family tree was his ability to take advantage of the rapid technological changes going on around him, either through his own initiative or through the suggestion of others. But Edward Pickering, as unique and gifted as he was (and would more than prove himself to be at the helm of the Harvard Observatory for more than forty years), was not unlike so many young men—and women, though fewer in number given prevailing social norms—who embodied the growing spirit of inquisitiveness, industry, and innovation that characterized much of nineteenth-century America, a century populated by reform-minded people willing to push the boundaries of traditional education. And push the boundaries they did with increasing speed, as they took advantage of the changing times, perhaps even helped fuel them, in order to explore the new frontier of hands-on, research-based, experiential learning in the ever-exciting but still emerging world of science and technology.

Significant Events

1802   Thomas Jefferson signs the Military Peace Establishment Act that establishes the U.S. Military Academy at West Point with the primary function of training engineers

1812   A group of naturalists interested in the distinctive character of American science forms the Academy of Natural Sciences in Philadelphia

1814   The Linnaean Society of New England forms in Boston by individuals interested in natural history; the group sponsors lectures, field trips, and a museum

1814   American scientist and adventurer Benjamin Thompson endows the Rumford Chair on the Application of Science to the Useful Arts at Harvard

1817   Sylvanus Thayer takes over as superintendent of the United States Military Academy, emphasizing both military leadership and civil engineering

1818   Benjamin Silliman, Sr. of Yale founds the *American Journal of Science and the Arts*, later shortened to the *American Journal of Science*, but just as often called "Silliman's *Journal*"

1821   The nation's first free public high school, the English High School, is founded in Boston with vocational course offerings

1823   Robert Hallowell Gardiner of Gardiner, Maine, founds the Gardiner Lyceum, the nation's first vocational trade school

1824   Samuel Merrick and William Keating establish the Franklin Institute in Philadelphia to promote the industrial and mechanic arts

1824   Stephen Van Rensselaer founds Rensselaer School in Troy, New York; it will evolve into the nation's first "true" polytechnic school, changing its name in 1861 to Rensselaer Polytechnic Institute

1825   Chartered in 1816, Thomas Jefferson's University of Virginia opens with eight separate schools of study: medicine, law, mathematics, chemistry, ancient languages, modern languages, natural philosophy, and moral philosophy

1826   The Maryland Institute for the Promotion of the Mechanic Arts is established in Baltimore modeled after Philadelphia's Franklin Institute

1826   Millbury Branch Number One in Millbury, Massachusetts, becomes the nation's first town lyceum based on the "mutual education" ideas of Josiah Holbrook

1826   Timothy Claxton forms the short-lived Boston Mechanics' Institution, reviving it several years later as the Boston Mechanics' Lyceum

1828   Under President Eliphalet Nott, Union College offers students a choice between two courses of study: liberal arts or scientific

1829   The Boston Society for the Diffusion of Useful Knowledge, also known as the Boston Lyceum, is founded

1830   Members of the defunct Linnaean Society reorganize as the Boston Society of Natural History, becoming one of Boston's premiere natural history societies

1836   Industrialist John Lowell, Jr. founds the Lowell Institute in Boston with a $250,000 gift; it will become the epicenter of the lyceum movement

1841   Harvard experiments with elective courses as alternatives to the traditional fixed curriculum at the urging of President Quincy and his science faculty

SIGNIFICANT EVENTS

1846   John C. Warren forms the "Warren Club" in Boston, later called the "Thursday Evening Club," to promote discussions of a social and scientific nature

1846   The Smithsonian Institution, funded by the estate of James Smithson, is formed for the purpose of disseminating knowledge of a scientific and practical nature

1847   Yale Corporation authorizes a fourth division or school, the Department of Philosophy and the Arts, focused on research and scholarship in applied chemistry

1847   Abbot Lawrence founds Lawrence Scientific School with a $50,000 gift to Harvard to promote engineering and industrial chemistry

1848   The American Association for the Advancement of Science is created from the narrowly focused Association of American Geologists and Naturalists

1850   Josiah Cooke, Erving Professor of Chemistry and Mineralogy at Harvard, establishes one of the nation's first teaching laboratories in chemistry

1852   New Hampshire native and Boston commission merchant Abiel Chandler endows the Chandler School of Science and the Arts at Dartmouth College

1852   Newly-elected president Henry Tappan models the University of Michigan's curriculum after the German "research model," expanding course offerings and encouraging the creation of research laboratories

1853   A group of New York businessmen establish the Brooklyn Collegiate and Polytechnic Institute that will evolve into the Polytechnic Institute of Brooklyn and later the Polytechnic Institute of New York University

1856 The Yale Scientific School, created two years earlier, changes its name to the Sheffield Scientific School in honor of major benefactor Joseph E. Sheffield

1859 Inventor, manufacturer, and businessman Peter Cooper establishes Cooper Union for the Advancement of Science and Art in New York City

1861 Yale becomes the first American institution to award the Doctor of Philosophy in an attempt to stem the flow of graduates going to Germany for advanced studies

1861 The Massachusetts legislature charters William Rogers' petition to establish an Institute of Technology in Boston's Back Bay; it will become the Massachusetts Institute of Technology (MIT)

1863 An elite group of American scientists led by Alexander Bache and Louis Agassiz, who call themselves the "Scientific Lazzaroni," create the National Academy of Sciences

1862 The U.S. Congress passes the Morrill Land-Grant College Act to provide incentives to states to establish programs in agriculture and the applied sciences

1864 Columbia College opens a School of Mines, the first such school in the nation; it will become a full-blown school of engineering within a matter of years

1865 The Worcester County Free Institute of Industrial Science is founded in Worcester, Massachusetts; it will evolve into Worcester Polytechnic Institute with the motto "Lehr und Kunst," translated "Theory and Practice"

1867 Edward Pickering, as assistant professor of physics at MIT, drafts "Plan of a Physical Laboratory" laying the groundwork for teaching physics in a laboratory environment; it will revolutionize laboratory teaching at other universities

SIGNIFICANT EVENTS

1867   After many years of military service, Sylvanus Thayer establishes the Thayer School of Engineering at his alma mater, Dartmouth College

1868   Taking advantage of the Morrill Land-Grant College Act, Andrew White and Ezra Cornell open Cornell University with an emphasis on applied science

1868   Newly-appointed president James McCook overhauls the College of New Jersey's curriculum, expanding course options in the natural and physical sciences; in 1896, the school will change its name to Princeton University

1869   Future Harvard president Charles Eliot publishes "The New Education" in the February/March issues of *The Atlantic Monthly* advocating for greater elective choices in the humanities and in the natural and physical sciences

1870   Stevens Institute of Technology opens in Hoboken, New Jersey, based on the European polytechnic model; it will offer programs in mechanical, civil, chemical, and electrical engineering

1872   The Commonwealth of Virginia charters the Virginia Agricultural and Mechanical College; it will evolve into Virginia Polytechnic Institute (VPI)

1876   Baltimore entrepreneur and philanthropist Johns Hopkins founds Johns Hopkins University; it is the first American university established as a research university from the outset

1881   Industrialist Edward Weston establishes the Newark Technical School with an emphasis on engineering; it will evolve into the New Jersey Institute of Technology

1884   The Jefferson Physical Laboratory opens at Harvard, becoming the first American university facility devoted exclusively to research and teaching in a scientific discipline

1885 Former California Gov. Leland Stanford and his wife, Jane Lathrop Stanford, endow Stanford University in memory of their son; the institution will be modeled after the "new" American research-based university

1887 Oil tycoon Charles Pratt founds Pratt Institute in the Clinton Hill section of Brooklyn with programs in engineering, architecture, and the fine arts in order to train industrial workers as engineers, mechanics, and technicians

1887 Businessman Jonas Gilman Clark endows Clark University in Worcester, Massachusetts, as the nation's first graduate studies institution offering programs in mathematics, physics, chemistry, biology, and psychology

1888 The Georgia School of Technology opens with an emphasis on applied science; it will evolve into Georgia Institute of Technology, a.k.a. Georgia Tech

1891 The Rochester Athenaeum, a literary society, and the Mechanics Institute of Rochester merge, laying the foundation for what will become the Rochester Institute of Technology

1891 Philadelphia financier Anthony Drexel founds Drexel Institute of Art, Science, and Industry with an emphasis on applied arts and science; it will become Drexel Institute of Technology and later Drexel University

1891 Amos Throop establishes a preparatory and vocational school in Pasadena, California that will evolve into the California Institute of Technology, the West Coast version of the Massachusetts Institute of Technology

Chapter Notes

CHAPTER 1 | *Land of Living Waters*

1. Thomas O'Connor, *Bibles, Brahmins and Bosses: A Short History of Boston* (Boston: Trustees of the Public Library of the City of Boston, 1984), 64.
2. Solon I. Bailey, "Biographical Memoir of Edward Charles Pickering, 1846-1919," *National Academy of Sciences Biographical Memoirs* 15 (1932): 169.
3. Harrison Ellery and Charles Pickering Bowditch, *The Pickering Genealogy: Being an Account of the First Three Generations of The Pickering Family of Salem, Mass, and of the Descendants of John and Sarah (Burrill) Pickering, of the Third Generation, Vols. 1-III* (Boston: Privately Printed, 1897).
4. John Pickering's family farm was located 30 miles northeast of Boston in Wenham, Massachusetts. Like several other area villages, Wenham had close ties to its southern neighbor Salem, one of the earliest and most important colonial seaports. Bordered by several large lakes and dense woodlands, Wenham's location—several miles east of Manchester-by-the-sea—and fertile land made it a prosperous farming community, first settled by the Algonquian peoples, a peaceful, agricultural group, and then in the mid-1630s by English colonists. Originally part of Salem, Wenham was incorporated in 1643 as an independent township with its own representative to the General Court. As with many early New England town names, Wenham, which means "home on the moor," probably got its name from early settlers who emigrated from Suffolk County, England, where there are two villages with the word Wenham in their name—Great Wenham and Little Wenham.
5. In 1806, John Pickering was asked to take one of two vacant professorships at Harvard, either the Hancock Professor of Hebrew and Oriental Languages or the Hollis Professor of Mathematics and Natural Philosophy. Regarding the former, it was said that Pickering, who had published widely on the nature of language, could read and write in more than twenty-two languages. Regarding the latter, although Pickering excelled in classical studies, he was equally at home in the world of mathematics, haven written a thesis in Latin on fluxions, the Newtonian form of calculus. Due to his more than full responsibilities, both legal and legislative, Pickering declined both offers. Eight years later, Pickering was offered the position of Eliot Professor of Greek Literature, but he declined that offer as well.

6. *Boston Directory, for the Year 1852, Embracing the City Record, A General Directory of the Citizens, and a Business Directory, with an Almanac, from July, 1852, to July, 1853* (Boston: George Adams, 1852).
7. Mary Orne Pickering, *Life of John Pickering* (Cambridge, MA: John Wilson & Son, 1887), 346.
8. Nancy Seasholes, the nation's authority on Boston's land-making projects, describes the original shape of the Shawmut Peninsula as a three-lobed leaf with coves separating each lobe. The coves are, moving counterclockwise from the southeast: South Cove, East or Great Cove, North Cove (latter Mill Pond), West Cove (a small marshy area northwest of the Common), and Back Bay (a large tidal bay bordering the Charles River). The other notable features of the peninsula, besides the slim "neck" of land that connects it to the mainland, are several hills or prominences projecting from the land: Fort Hill overlooking the Great Cove, Copp's Hill in the North End, and Trimountain, or Trimount (from which Tremont Street is derived) running east to west across the middle of the peninsula with its three distinct prominences—Mount Vernon, Beacon or Sentry Hill, and Pemberton or Cotton Hill (along with a smaller knob called Fox Hill on the western edge of the Common).
9. Marc Anthony de Wolfe Howe, *Boston Common: Scenes from Four Centuries* (Cambridge, MA: The Riverside Press, 1910), 10-11.
10. Harrison Gray Otis was one of the prime movers behind many of the land development projects undertaken in Boston in the early decades of the nineteenth century. The nephew of patriot James Otis and grandson of Harrison Otis, the last Tory treasurer of the Massachusetts Bay Colony, Otis should have started life with a silver spoon in his mouth, but after the Commonwealth of Massachusetts confiscated his family's estate, he began his legal career as a man of modest means. But it didn't take long for the intelligent, perceptive, and ambitious lawyer to climb Boston's social and economic ladder. It was while serving on the committee named in 1794 to select a site for the New State House that Otis, along with fellow lawyer and legislator Jonathan Mason, got the idea to develop land west of Beacon Hill. At the time, the area was nothing more than a thicket of brambles crisscrossed by a handful of cow paths. Along with several other investors, including Charles Bulfinch, Otis and Mason formed the Mount Vernon Proprietors.
11. If anyone facilitated the merging of architect and housewright in the boom years following the War of 1812, it was Connecticut-born Asher Benjamin, a talented architect and master builder, who published several architectural "pattern books" with the intent of introducing architectural history, style, and geometry to master craftsmen in the field. Although first influenced by Bulfinch's urbane Federal style (Benjamin not only worked on Hartford's Old State House, one of Bulfinch's earliest building projects, but also lived in Boston during the height

of Bulfinch's reign), Benjamin shifted his tastes toward Greek-Revival architecture later in his career, with many of his designs adapted from the work of British architects James Gibbs and Colen Campbell.
12. *Images of America: Boston Common* (Charleston, SC: Arcadia Publishing, 2005), 28.
13. Michael Rawson, *Eden on the Charles: The Making of Boston* (Cambridge, MA: Harvard University Press, 2010), 31.
14. Digital Commonwealth: Massachusetts Collections Online. Boston Public Library, https://www.digitalcommonwealth.org/search/commonwealth:4q77fw147.
15. Rawson, *Eden on the Charles*, 34.
16. A. McVoy McIntyre, *Beacon Hill: A Walking Tour* (Boston: Little, Brown & Co., 1975), 15.
17. As far-fetched as it seems, it is true: according to Christopher Klein, writing for the History Channel, Frederic Tudor (1783-1864), founder of the Tudor Ice Company, pioneered the international ice trade in the early 19th century, shipping New England ice, cut from frozen winter ponds (including Thoreau's coveted pond outside of Concord), to countries as far away as India and Brazil. And he was only twenty-two years old when he concocted the scheme, leading skeptics, including his father, to scoff at the idea. Undeterred, Tudor bought his own ship and set sail for Martinique with a hold full of ice covered with straw. Although he encountered setbacks at every turn, Tudor pushed ahead, learning from his mistakes, until finally, after the invention of a revolutionary two-bladed horse-drawn ice cutter (attributed to Nathaniel Wyeth, one of his suppliers and eventual business partner) that decreased the time and effort required to extract ice from New England ponds, Tudor began to see his vision realized and in the course of time dominated the international ice trade making him a very wealthy man. Source: Christopher Klein, "The Man Who Shipped New England Ice Around the World," *The History Channel*, www.history.com/news/the-man-who-shipped-new-england-ice-around-the-world.

CHAPTER 2 | *Reading, Writing, Refmetic*

1. James D'Wolf Lovett, *Old Boston Boys and the Games They Played* (Boston: Little, Brown & Co., 1908), 3.
2. Old Deluder Satan Act, New England Historical Society, http://www.newenglandhistoricalsociety.com/old-deluder-satan-act-made-sure-puritan-children-got-educated.
3. Malden Public Library, Malden, Massachusetts, *Local History*, http://maldenpubliclibrary.org/browse-mpl/local-history.
4. The full quotation can be found in Andrew Rickoff, *Past and Present of Our Common School Education: Reply to President B. A. Hinsdale, with

a Brief Sketch of the History of Elementary Education in America (Cleveland: Leader Print Co., 1877), 10; and Alexander Inglis, *The Rise of the High School in Massachusetts* (New York: Teachers College, Columbia University, 1911), 26-27.
5. The full quotation can be found in Andrew Rickoff, *Past and Present*, 8; and George Gary Bush, "History of Higher Education in Massachusetts," in *Contributions to American Educational History*, edited by Herbert B. Adams (Washington, DC: Bureau of Education Circular of Information, 1891), 24.
6. Horace Mann, *Twelfth Annual Report of the Board of Education, together with the Twelfth Annual Report of the Secretary of the Board (1848)*, State Library of Massachusetts, http://archives.lib.state.ma.us/handle/2452/204731.
7. Ellwood Cubberley, *The History of Education: Educational Practice and Progress Considered as a Phase of the Development and Spread of Western Civilization* (Cambridge: The Riverside Press, 1920), 690. For another view of the educational reform during the early nineteenth century, especially related to initiatives in northern states, see Thomas F. Army, Jr., "Common School Reform and Science Education." *Engineering Victory: How Technology Won the Civil War* (Baltimore: Johns Hopkins University Press, 2016).
8. Lovett, *Old Boston Boys*, 3.
9. Ibid., 8.
10. Ibid., 81.
11. Ernest Samuels, *Henry Adams* (Cambridge: The Belknap Press of Harvard University, 1995), 7.
12. Henry Cabot Lodge, *Early Memories* (New York: Charles Scribner's Sons, 1913), 81.
13. Ibid., 82-83.

CHAPTER 3 | *Schools and Schoolmasters*

1. George Santayana, *The News-Palladium* (April 29, 1935), 2.
2. For an overview of the early days of Boston Latin School, see Henry Jenks, *The Boston Public Latin School, 1635-1880* (Cambridge, MA: Moses King, 1881) and Pauline Holmes, *A Tercentenary History of the Boston Public Latin School, 1635-1935* (Cambridge, MA: Harvard University Press, 1935).
3. Holmes, *A Tercentenary History*, 27.
4. Ibid., 235.
5. Although we attribute to Ezekiel Cheever the work titled *Cheever's Accidence*, there is still some question as to whether or not Cheever was the original author. Pauline Holmes suggests that he may not have been. Citing George Littlefield, author of *Early Schools and School-*

Books of New England, Holmes suggests that the author of the 1709 edition (printed one year after Ezekiel Cheever's death) may have been the work of either Cheever's son, the Rev. Thomas Cheever, Cheever's grandson, Ezekiel Lewis, or Nathaniel Williams, who succeeded Cheever as the Master of Boston Latin in 1708. On the other hand, noted educator Henry Barnard, who edited the *American Journal of Education*, at the time the nation's most prestigious publication about American education, does not waver in his assessment of who wrote *Cheever's Accidence*—Ezekiel Cheever, that's who, noting that by 1790 the *Accidence*, which carried the subtitle *A Short Introduction to the Latin Tongue*, had "passed through twenty editions, and was for more than a century the hand-book of most of the Latin scholars of New England." This discussion can be found in *American Journal of Education* 12 (1862), 545.

6. *Catalogue of the Boston Public Latin School, Established in 1635, with an Historical Sketch Prepared by Henry F. Jenks* (Boston: The Boston Latin School Association, 1886), 28.
7. It is no hyperbole to use the term "affectionately" when speaking of Boston Latin, as reflected in Lee Daniels' reminiscence of his time at the school on the 350th anniversary of Boston Latin: "Recently, when I returned to my secondary school, the Boston Latin School, for the first time in years, I initially thought I had stepped into a time warp. For, as I strolled its corridors and looked into its classrooms, I was immediately swept up in seven years' worth of memories—of knowledge and values learned, of failure turned to triumph, of friendships formed. I walked into its empty auditorium and, as when I was a student, the very air seemed to resound with names from its three-century-long alumni roster: Cotton Mather, Sam Adams, John Hancock, Ralph Waldo Emerson, George Santayana, Bernard Berenson, Leonard Bernstein. I felt I stood on hallowed ground, and it was wonderful to be there again." Source: Lee Daniels, "The Halls of Boston Latin School," *The New York Times* (April 21, 1985), https://www.nytimes.com/1985/04/21/magazine/the-halls-of-boston-latin-school.html.
8. Ellen Chase, *The Beginnings of the American Revolution: Based on Contemporary Letters, Diaries, and Other Documents*, vol. 3 (New York: The Baker and Taylor Company, 1910), 101.
9. William R. Dimmock, *Memorial of Francis Gardner, LL.D., Late Head-Master of the Boston Latin School* (Boston: The Boston Latin School Association, 1876), 38.
10. Phillips Brooks, *Essays and Addresses: Religious, Literary and Social*, vol. 3 (New York: E. P. Dutton & Company, 1894), 419.
11. The pedagogy of the liberal arts curriculum in most American colleges prior to the Civil War revolved around three cornerstone pratices: recitations, declamations, and disputations. The pedagogy of recitations

required that students memorize a few pages assigned from a textbook prior to each day's class and, when called upon during class, repeat aloud the assigned section of the text. It was an age-old practice established to discipline the mind, a rigid form of character development intended to strenghen the mental powers of future leaders—doctors, lawyers, judges, teachers, and ministers. Declamation, on the other hand, was a public speaking exercise held once or twice a week, often in the college chapel. One or more students from each class made a public address or declamation before the faculty and student body, with the college president and professor of rhetoric critiquing each speaker after his performance on both accuracy and presentation. Like class recitations, declamations were memorized, but with the subject matter selected by the student, not the instructor. Disputations, as the term implies, involves the defence of a thesis or position. Rather than rote memorization, what was called for here was the development a logical argument based on evidentiary propositions. As such, it was more syllogistic than rhetorical. Colleges took very seriously these three cornerstones of pedagogy as the following excerpt indicates: "On Tuesdays and Fridays every undergraduate in his turn, about six at a time, shall declaim in the English, Latin, Greek or Hebrew Tongue, and in no other, without special liberty from the President, and shall presently after deliver up his declamation to his Tutor, fairly written, with his name subscribed. The two senior classes shall dispute twice a week in the chapel; and if any undergraduate shall be absent from recitation or disputation without liberty, he may be fined two pence, and if from declaiming six pence." Source: "Yale College, Concerning Scholastic Exercises," in *The Liberal Arts Tradition: A Documentary History*, edited by Bruce A. Kimball (Lanham, MD: University Press of America, 2010), 212.
12. Holmes, *A Tercentenary History*, 264.
13. Ibid., 275.

CHAPTER 4 | *The Leaven of Improvement*

1. Samuel Rezneck, *Education for a Technological Society: A Sesquicentennial History of Rensselaer Polytechnic Institute* (Troy, NY: Rensselaer Polytechnic Institute, 1968), 17.
2. Frederick Rudolph, *The American College and University: A History* (Athens, GA: The University of Georgia Press, 1990), 49. Boston was not the only city in the new republic that identified with the ancient Greek world. As Rudolph notes, the Greek revival in America in the latter eighteenth and early nineteenth centuries "would not only dot the landscape with adaptations of classical architecture but it would give almost every state its aspiring Athens." When trustees of the College of

New Jersey in Princeton appealed to the state legislature for financial aid in 1795, they did so, declaring, "We will make New Jersey the Athens of America." For another discussion on the origin of this reference, see Robert Keogh's article in the spring 2006 issue of *Commonwealth: Politics, Ideas & Civic Life in Massachusetts*, titled "Historian Thomas O'Connor on Making Boston the Athens of America."

3. *Annals of the Massachusetts Charitable Mechanic Association, 1795-1892* (Boston: Rockwell & Churchill, 1892), 1.
4. Arthur Wellington Brayley, *Schools and Schoolboys of Old Boston* (Boston: Louis P. Hager, 1894), 99.
5. *The American Journal of Education* 8, no. 20 (March 1860): 230-232. The content of Holbrook's letter is contained within a larger discussion of Holbrook's life and impact on the direction of the American Lyceum Movement.
6. Angela Ray, *Lyceum and Public Culture in the Nineteenth-Century United States* (East Lansing: Michigan State University Press, 2005), 5.
7. Timothy Claxton, *Memoir of a Mechanic: Being a Sketch of the Life of Timothy Claxton, Written by Himself. Together With Miscellaneous Papers* (Boston: George W. Light Publisher, 1839), 6.
8. John Pickering was a classical scholar and philologist who twice turned down appointments to teach at Harvard. A lawyer by training, Pickering moved his law practice to Boston in 1827, the year after he published an important lexicon of the Greek language. A civic-minded person, Pickering became involved in local politics, serving as city alderman and chairman of the committee regulating the Boston Latin School. In 1829, he became city solicitor of Boston, vice president of the Boston Society for the Diffusion of Useful Knowledge (publishing the society's first annual report in 1830), and later represented Suffolk County in the Massachusetts State Senate.
9. Helen Deese and Guy Woodall, "A Calendar of Lectures Presented by the Boston Society for the Diffusion of Useful Knowledge, 1829-1847," *Studies in the American Renaissance* (Charlottesville, VA: The University of Virginia Press, 1986), 17-67.
10. John Lowell, *Extracts from the Will and Codicil of John Lowell, Jr, Concerning the Lecture Fund and Opinions of Hon. Chas. G. Loring, and Hon. Benj. R. Curtis Upon the Duties of the Trustees of the Boston Athenaeum as Visitors of the Lecture Fund* (Boston: James F. Cotter & Co., Printers, 1885), 9.
11. Thomas Jefferson did not create the Corps of Engineers; credit goes to the Continental Congress when it organized the Continental Army in the summer of 1775, authorizing a chief engineer and two assistants. Washington chose Richard Gridley as his first chief engineer, charging him with the task of building fortifications near Boston at Bunker Hill. Beyond Gridley, there were very few engineers trained in military forti-

fications living in the colonies, so the Continental Congress turned to the government of King Louis XVI of France for help. When the Continental Congress created a separate Corps of Engineers in the spring of 1779, it was not Gridley or any other colonist who took charge of the unit. It was Louis Duportail, a member of the French Royal Corps of Engineers, who came to the colonies secretly two years earlier to help Washington and the Continental Army and was instrumental in directing the construction of siege works at the decisive Battle of Yorktown.

12. Napoleon, two-time emperor of the French nation, showed an early interest in science, firmly believing that advances in science and technology were fundamental to the wellbeing of the French people. Trained as a military engineer with formidable mathematical skills, Napoleon was elected to the National Institute, the foremost scientific society in post Revolutionary France, in 1797. Napoleon's motto—*Pour la Patrie, les Sciences et la Gloire*, literally "For the Homeland or Nation, Science and Glory"—rang especially true the following year during the invasion of Egypt in which Napoleon, wanting to bring to the founding country of western culture the "gift" of science and technology, brought with him a corps of scholars, trained in astronomy, natural history, linguistics, topography, and engineering.

13. West Point's motto—Duty, Honor, Country—was officially adopted in 1898, the year West Point's coat of arms was designed. The principal imagery of the coat of arms includes the helmet of Pallas Athena, a fully-armed mythological goddess associated with the art of war, attached to a shield bearing the colors of the United States, and a bald eagle sitting atop the shield, clutching thirteen arrows (representing the 13 original colonies), oak and olive branches, and a scroll that bears the phrase "West Point, MDCCCII, USMA" and the motto "Duty, Honor, Country." Although these words had surfaced independently in the writings of West Point administrators and faculty prior the creation of West Point's coat of arms (especially in Thayer's description of his "triangle" of honor, discipline, and education), 1898 is the first year that "duty," "honor," and "country" appear together as the official motto of the U.S. Military Academy.

14. Lawrence Grayson, *The Making of an Engineer: An Illustrated History of Engineering Education in the United States and Canada* (New York: John Wiley & Sons, 1993), 23.

15. Thayer's resignation in 1833 didn't end his influence at West Point. Between 1833 and 1871, Dennis Hart Mahan, senior professor of engineering at the military academy, embraced Thayer's curriculum and instructional methods, solidifying the cadet course as a curriculum devoted to engineering. Mahan, who graduated at the top of his West Point class in 1824, rose through the faculty ranks to become head of the Engineering Department, and for almost 50 years taught military

science, fortifications, and civil engineering. He firmly believed Thayer's proposition: that the study of engineering was the proper and necessary means to produce competent military officers, who could both fortify the nation and lead United States soldiers into battle. Noting the importance of the engineering curriculum at West Point under Thayer and Mahan, John Scot Logel writes: "While the engineering success of an Academy graduate was not preordained upon completing West Point, the foundation in engineering enabled those who did succeed to do so immeasurably. Graduates of the Military Academy used various elements of engineering theory from Mahan's courses to build railroads, survey and map the topography of the United States, establish new schools for engineering, and contribute to the professionalism of the engineering field. West Point's engineering curriculum was the foundation for engineering education and improvement in the United States." Source: Jon Scott Logel, *Designing Gotham: West Point Engineers and the Rise of Modern New York, 1817-1898* (Baton Rouge: Louisiana State University Press, 2016), 53-54.
16. Rezneck, *Education for a Technological Society*, 92.

CHAPTER 5 | *"Old Sheff"*

1. Alfred E. Walker, "Show Me the Way to Old Yale, Boys," in *Trigintennial Record of the Class of 1876 of the Sheffield Scientific School of Yale College* (New Haven: Tuttle, Morehouse & Taylor, 1908), 43.
2. Rudolph, *The American College and University*, 25-26.
3. Thomas Denham, "A Historical Review of Curriculum in American Higher Education: 1636-1900," (ERIC DOCUMENT 471739), 3.
4. Ibid., 7.
5. Ibid., 7.
6. Julie Kern, "The Yale Report of 1828: A Synopsis," https://www3.nd.edu/~rbarger/www7/yalerpt.html.
7. Jonathan Lyons, *The Society of Useful Knowledge: How Benjamin Franklin and Friends Brought the Enlightenment to America* (New York: Bloomsbury Press, 2013), 105.
8. Rudolph, *The American College and University*, 31-32.
9. *Reports on the Course of Instruction in Yale College; by a Committee of the Corporation, and the Academical Faculty* (New Haven: Yale Corporation, 1828), 6.
10. Ibid., 7.
11. Ibid., 7-8.
12. Ibid., 8.
13. Ibid., 8.
14. Ibid., 14.
15. Ibid., 16.

16. John Fulton and Elizabeth Thompson, *Benjamin Silliman: Pathfinder in American Science* (New York: Henry Schuman, 1947), 10.
17. Chandos Michael Brown, *Benjamin Silliman: A Life in the Young Republic* (Princeton, NJ: Princeton University Press, 1989), 55.
18. Ibid., 57.
19. Ibid., 83.
20. Among his other achievements, most of them accomplished after 1828, Benjamin Silliman, Sr. helped Yale establish the first art gallery sponsored by an institution of higher learning after he convinced his wife's uncle, Col. John Trumbull, to donate a number of his paintings to Yale. Silliman was one of the most sought-after speakers on the lyceum circuit (evidenced not only by his many wide-ranging lecture engagements, but also by giving the inaugural lecture at the opening of the Lowell Institute in Boston). And, as we shall see, he—and his son, Benjamin Silliman, Jr.—was the motivating force behind the establishment of the Sheffield Scientific School, one of the first schools in America focused on science education.
21. Brown, *Benjamin Silliman*, 105.
22. Noted teacher, scholar, entrepreneur, inventor, and preacher, Eliphalet Nott was the longest serving president of Union College (1804-1866), perched above the Mohawk River in Schenectady, New York. Its location afforded Nott another position: president of Rensselaer Institute from 1829 to 1845. One man, two presidencies: it worked because Nott's title at Rensselaer was more in name than substance (he visited Rensselaer once a month to consult with its trustees and headmaster). A graduate of Rhode Island College, Nott's presence—even if just in name—lent credibility to Rensselaer, given his many connections throughout the state, forged through his time as minister of the First Presbyterian Church of Albany and chaplain of the state legislature. In 1804, Nott succeeded Rev. John Blair Smith as Union College's fourth president, exerting a strong hand in its governance and curriculum development. Perhaps his most significant innovation was to promote the parity of classical and practical education. To this extent, Nott instituted one of the nation's first parallel curricula enabling students to choose between ancient and modern languages, and between historical subjects and practical technology, with all students receiving the same degree—a bachelor of arts—upon graduation.
23. Charles H. Warren, "The Sheffield Scientific School from 1847 to 1947," in *The Centennial of the Sheffield Scientific School*, edited by George R. Baitsell (New Haven: Yale University Press, 1950), 156.
24. With little or no aptitude for the classics, John Pitkin Norton preferred doing chores on his father's farm to attending classes in Latin and Greek language and literature. He did study, but primarily as a private student, first at Yale during with winter of 1840-1841 attending

lectures on chemistry and mineralogy, then at Harvard the following winter studying chemistry and anatomy. He returned to Yale for additional studies between 1842-1844, whereupon he traveled to Scotland to study agricultural chemistry in the laboratory of the Agricultural Chemical Association in Edinburgh, one of the most progressive laboratories in the Western world. After he accepted the professorship in agricultural chemistry and physiology at Yale in the summer of 1846, Norton returned to Europe for nine months to study at a laboratory in Utrecht to further prepare for his new position at Yale. Norton is not unusual in this way: many professors studied at European universities, especially in Germany, to "finish" their education, since European science education was far superior to what they could find in America.

25. Fulton and Thomson. *Benjamin Silliman,* 211.
26. Ibid., 212-213.
27. Warren, "The Sheffield Scientific School," 158.
28. John Whitehead, *The Separation of College and State: Columbia, Dartmouth, Harvard, and Yale, 1776-1876* (New Haven: Yale University Press, 1973), 172.
29. Instrumental in the redesign of the Sheffield Scientific School at Yale was Daniel Coit Gilman, Yale Class of 1852, who served as librarian of Yale College from 1856 until the outbreak of the American Civil War. In 1863, Gilman held the title Professor of Physical and Political Geography at the Sheffield Scientific School, becoming the school's secretary and librarian several years later. In 1872, he resigned his responsibilities at Sheffield Scientific to assume the presidency of the University of California. Three years later, he resigned that position to become the first president of Johns Hopkins University, the first research-based graduate school in America. In a tribute to the Sheffield Scientific School celebrating its fiftieth anniversary in 1897, Gilman wrote this about the impact "Old Sheff" had upon his thinking: "By this course of remarks you have been reminded that this school was founded in favorable environs, at a propitious time, and also that it is only one of many kindred agencies initiated within the period under review. The Lawrence Scientific School of Harvard was almost coeval. In quick succession, colleges, departments of science and independent institutions have appeared in every state. Of these not a few have adopted the methods here followed or have called to their support those who have been trained here. For one such institution, now celebrating its majority, permit me to acknowledge with filial gratitude the impulses, lessons, warnings, and encouragements derived from the Sheffield Scientific School, and publicly admit that much of the health and strength of the Johns Hopkins University is due to early and repeated draughts upon the life-giving springs of New Haven." Source: Daniel Coit Gilman,

The Sheffield Scientific School of Yale: A Semi-Centennial Historical Discourse (New Haven: Sheffield Scientific School, 1897), 37.
30. Leonard G. Wilson, "Benjamin Silliman: A Biographical Sketch," in *Benjamin Silliman and His Circle: Studies on the Influence of Benjamin Silliman on Science in America*, edited by Leonard G. Wilson (New York: Science History Publications, 1979), 8-9.

CHAPTER 6 | *An Impenetrable Thicket*

1. George Ticknor, *Life, Letters, and Journals of George Ticknor in Two Volumes* (Boston: James R. Osgood and Company, 1877), I:359.
2. In 1584, Sir Walter Mildmay founded Emmanuel College in Cambridge, England. The mission of the college was to train preaching ministers and then to send them out in every direction to fill the parishes and pulpits of churches in rural England. Founded on the strict principles of Calvinist Protestantism, Emmanuel College became not only a leader in training orthodox Puritan ministers, but also one of the largest colleges in Cambridge during the early seventeenth century. It was also responsible for the education of many of England's "dissenting" clergy, including the Rev. John Cotton, one of the earliest Puritan clergy to immigrate to the New World.
3. Samuel Morison, *Three Centuries Harvard History* (Cambridge, MA: The Belknap Press of Harvard University, 1965), 12.
4. "John Leverett, 1662/3-1724," in *The Bloomsbury Encyclopedia of American Enlightenment*, edited by Mark Spencer (New York & London: Bloomsbury Academic, 2015), 2:624.
5. Fire was an ever-present danger in an age of wood-frame structures. After the fire that decimated Harvard Hall, the following regulations were adopted for the new library building. First of all, the building's walls were made of brick and stone, not wood; and its roof made of slate, not wooden shingles. Secondly, since the building continued to house the college library, no student dormitories were placed in the building because the use of fireplaces and lamps by students at night posed a fire hazard. Moreover, no patron could bring a candle or lamp into the library. Finally, if the library's fireplace was lit, a staff member had to be present at all times to make sure no embers escaped the hearth and that the fire was properly extinguished at the end of the night.
6. The term "Awakening" refers to the cyclical appearance of an evangelical revival that often correlates with the ebb and flow of moral values. The First Great Awakening swept through the British Isles and the American colonies between the 1730s and 1740s. Although primarily a Protestant movement, the revival was pan-denominational in nature as its leaders—often uneducated, itinerant preachers—attempted to build

a universal or common evangelical identity among their followers. At the core of any revival or "awakening" is religious conversion, a "born-again" spirit of renewed faith predicated upon the deep conviction of the need for salvation. Since the Second Great Awakening began on the cusp of a new millennium, peaking in America in the 1830s, many of its adherents believed that it heralded a new millennial age, presaging the Second Coming of Christ. As such, it was a highly charged, emotional experience that rejected the rationalism of the European Enlightenment.

7. When referring to Harvard, I often use the term "Harvard College," which today refers exclusively to the four-year undergraduate program. Although the Massachusetts Constitution officially recognized Harvard as a university in 1780, I prefer the term Harvard College (even after the 1780 date), since I'm usually referring to the traditional or "classical" undergraduate liberal arts program.
8. Robert McCaughey, *Josiah Quincy, 1772-1864: The Last Federalist* (Cambridge, MA: Harvard University Press, 1974), 147.
9. Ibid., 156.
10. Student uprisings were not exclusive to Harvard during the antebellum period. At Yale, as Corydon Ireland notes in an article for *The Harvard Gazette*, things were much rougher. Between 1843 and 1858, students carried arms, stabbed tutors, built barricades, and engaged by the hundreds in town brawls (with one resulting in the death of two New Haven firemen). Ireland offers several reasons for the rowdy, even criminal, behavior of college students: a growing disenchantment with college authority and the lack of choice in the undergraduate curriculum; the arrival of older students on college campuses; and the proliferation of social clubs with its excesses of drinking and all-night carousing. At Harvard, where students often congregated around "Rebellion Tree" in Harvard Yard, the era of rowdiness came to a screeching halt in 1829 with the appointment of Josiah Quincy as Harvard's seventeenth president. The hardheaded, no-nonsense judge, a veteran of the U.S. House of Representatives and six-term mayor of Boston, vowed to bring law and order to the college and did so, often lodging civil and criminal charges against rebellious students. When he retired in 1845, as Ireland notes, Quincy had doubled the endowment, introduced a system of elective courses, and elevated the college to university status, but increased personal responsibility and discipline among students may have been his greatest achievement.
11. McCaughey, *Josiah Quincy*, 171.
12. President Josiah Quincy's introduction of elective courses at Harvard would not last, until Charles Eliot became Harvard's twenty-first president in 1869. Holding one of the longest tenures in that office, 1869-1909, Eliot would institute an elective system beginning in the

freshman year, to which many balked, believing that eighteen- and nineteen-year-olds were not mature enough to handle such a process.
13. John Todd, *Death in the Palace: A Sermon in Memory of Edward Everett, January 22, 1865* (Boston: Dakin & Metcalf, Printers, 1865), 10.
14. Paul Varg, *Edward Everett: The Intellectual in the Turmoil of Politics* (Toronto: Associated University Presses, 1992), 127.
15. Edward Everett, "Inaugural Address," *Addresses at the Inauguration of the Hon. Edward Everett, LL.D., as President of the University at Cambridge, Thursday, April 30, 1846* (Boston: Charles C. Little & James Brown, 1846), 34.
16. Ibid., 34.
17. Ibid., 43.
18. Ibid., 48.
19. Ibid., 53.
20. Ibid., 55.
21. The official title of the Rumford endowed chair is Rumford Professor on the Application of Science to the Useful Arts, a somewhat archaic and tedious title, hence the shorthand: Rumford Professor of Applied Science.
22. It is no hollow saying that Benjamin Thompson never forgot his roots: born in Rumford, Massachusetts, Thompson chose his title carefully, Count Rumford, which honored both his native country and his place of birth.
23. Mary Ann James, "Engineering an Environment for Change: Bigelow, Peirce, and Early Nineteenth-Century Practical Education at Harvard," in *Science at Harvard University: Historical Perspectives*, edited by Clark A. Elliott and Margaret W. Rossiter (Bethlehem, PA: Lehigh University Press, 1992), 60.
24. Ibid., 60.
25. Ibid., 62.

CHAPTER 7 | *Bridge to the Future*

1. *Twenty-Second Annual Report of the President of the University at Cambridge to the Overseers, Exhibiting the State of the Institution for the Academical Year 1846-47* (Cambridge, MA: Harvard University Press, 1848), 24.
2. Clark Elliott, "Founding of the Lawrence Scientific School at Harvard University, 1846-1847: A Study in Writing and History," *Archivaria* 38 (1994): 121.
3. *Ibid.*, 123.
4. Hamilton Andrews Hill, *Memoir of Abbott Lawrence* (Boston: Little, Brown, and Company, 1884), 117.
5. *Ibid.*, 120.

6. *Ibid.*, 120.
7. *Ibid.*, 121.
8. *Ibid.*, 121-122.
9. Rudolph, *The American College and University*, 179.
10. If Edward Pickering were the son of a Boston Brahmin family with unlimited funds at its disposal, he would have had several other options for travel: he could take the family chaise, a two- or four-seat horse and buggy; if his family didn't own a chaise, he could hire a private chaise to take him the distance; or he could take a horse-drawn omnibus, the horse-drawn streetcar's main competitor. Unlike the omnibus that could venture anywhere (and, as such, was more expensive to take), the streetcar was restricted to thoroughfares with set tracks laid down in the middle of the street. When the Cambridge Horse Railroad opened in the spring of 1856, the first of its kind in Boston, the horse-drawn streetcar ran from Bowdoin Square in Boston to Central Square in Cambridge, later expanding service on both ends of the line's main route. Within a short period of time, four horse-drawn streetcar lines emerged to dominate the greater Boston area, each maintaining a monopoly over its service routes.
11. Bainbridge Bunting, *Harvard: An Architectural History*, completed and edited by Margaret Henderson Floyd (Cambridge, MA: Belknap Press of Harvard University, 1985), 36.
12. A curious family that even today is a household name for Harvard graduates, since most of them used HOLLIS, Harvard's On-Line Library Information System, to complete their assignments.
13. Eben Horsford's most successful product is still today a household name: Rumford Baking Powder, which was named after Horsford's official title at Harvard, the somewhat archaic and tedious—as I've already mentioned—Rumford Professor on the Application of Science to the Useful Arts.
14. Stanley M. Guralnick, *Science and the Ante-Bellum American College* (Philadelphia: American Philosophical Society, 1975), 64-65.
15. "Memorial of Joseph Lovering, Late President of the Academy," *American Academy of Arts and Sciences* (Cambridge, MA: John Wilson and Son, 1892), 15.
16. *Ibid.*, 15.
17. Although John Farrar does not take center stage in this study, he was—like his colleague Benjamin Peirce—a giant in the field of mathematics. A Harvard alumnus (Class of 1803), Farrar began his academic career as a Greek tutor at his alma mater in 1805. Two years later, he was appointed Hollis Professor of Mathematics and Natural Philosophy, a position for which he was well suited, as he had many interests. He studied weather patterns in the greater Boston area, conducted a detailed analysis of the hurricane that hit New England

in 1815 (describing for the first time a storm of such magnitude as having a "moving vortex"), and he attempted on several occasions to establish an astronomical observatory at Harvard. He is remembered most, however, for his work in the field of mathematics, specifically for his translations of mathematical treatises from European scholars and his creation of the Cambridge mathematical series, which became the standard for teaching mathematics at Harvard and other universities for years to come.

18. Florian Cajori, *A History of Physics in its Elementary Branches Including the Evolution of Physical Laboratories* (New York: MacMillan Company, 1922), 287.

19. The Flemish painter David Teniers the Younger (1610-1690) painted one of the earliest images of a chemistry laboratory. Of course it is not the chemistry laboratory that you and I are familiar with: it is the laboratory of an alchemist. Titled "An Alchemist in his Workshop" (a subject his father, Teniers the Elder, had painted before him, though with less mastery), Teniers depicts a bearded, gray-haired man intently at work, stoking a fire beneath an oblong crucible with a hand bellows. Hammers, tongs, vials, and urns (as well as several thick books) surround him as he works undisturbed. Teniers's portrait reminds us that the first laboratory was that of the alchemist, not the chemistry professor; but more than that, it reminds us that the first chemistry laboratories were often in an investigator's kitchen and not in the basement of a university building.

20. Along with their different views on the meaning and importance of Darwin's theory of evolution through natural selection, Agassiz and Rogers differed on the nature and role of science in an institution of higher learning, with Agassiz favoring pure or basic research and Rogers clinging to the idea of useful or applied science. When it comes to pure research, the emphasis is on curiosity, which is typically driven by some unanswered question or unsolved riddle that perplexes a scientist. It is, in short, the pursuit of knowledge for knowledge's sake rather than, as in the case of applied science, the solving of a practical problem that often leads to the creation or invention of something. For pure scientists, the concern is the continuous expansion of knowledge regardless of the practical applications of scientific findings; for applied scientists, the concern is the improvement of the human condition through a results- or outcome-driven process. In the early days of Harvard's Lawrence Scientific School and the Massachusetts Institute of Technology, the differences in scientific attitude held by Agassiz and Rogers respectively contributed greatly to the different directions each school took both in hiring new faculty and in shaping its curriculum.

21. One name does stand out among the thirteen students majoring in physics, at least for the purposes of this study. But a second name is

worth noting—John Trowbridge. Although three years older than Edward Pickering, Trowbridge was Pickering's classmate at the Lawrence Scientific School, graduating with Pickering in 1865 with a degree in physics. However, they were more than classmates; they were each other's alter egos. In a tribute to Trowbridge written for the National Academy of Sciences, physicist Edwin H. Hall observed the following: "It is my surmise that [Trowbridge's] career may have been determined to some extent by that of his contemporary and friend Edward C. Pickering, whose natural bent for science was unmistakable. The two graduated from the Lawrence Scientific School of Harvard University in the same year, 1865, with the degree of S.B. [Bachelor of Science]. Each taught mathematics for two or three years at Harvard; each was afterward for a time in the physics department of the Massachusetts Institute of Technology; each returned finally to the service of Harvard, Trowbridge as assistant professor of physics in 1870, Pickering as professor of astronomy in 1876; for about forty years they were near neighbors in Cambridge. Pickering was, I think, the first American to write a laboratory manual of physics. Trowbridge was one of the first, if not the very first, of Americans to put students into the way of original research in physics." Source: Edwin Hall, "Biographical Memoir of John Trowbridge, 1843-1923," *National Academy of Sciences Biographical Memoirs* 14 (1930): 186-187.

CHAPTER 8 | *Outshine Them All*

1. William Barton Rogers, *Life and Letters of William Barton Rogers*. Edited by Emma Savage Rogers with the assistance of William T. Sedgwick, in two volumes (Boston & New York: Houghton, Mifflin & Co., 1896), 1:311.
2. Prescott's full citation is worth noting: "The career of the Rogers brothers coincided with the rapid development of science, technology, and industry in the United States in the first half of the nineteenth century. In their early manhood while teaching in Baltimore, they had shared the popular excitement over the projected canal and railway. They had participated in geological surveys that supplied the basic information for mines and railways throughout the East, and they had witnessed the beginnings of large-scale industry in New England. With their interest in science and in education, it was natural that they should sense the need for a new kind of education to serve as the basis for the great industrial and technological development that they foresaw. It was also natural that they should choose Boston as the most promising location for an institute of technology." Source: Samuel Prescott, *When M.I.T. was "Boston Tech," 1861-1916* (Cambridge, MA: The Massachusetts Institute of Technology Press, 1954), 21.

3. The University of Pennsylvania emerged out of the shadows of two previous institutions: Benjamin Franklin's Academy of Philadelphia and William Smith's College of Philadelphia. Franklin founded the Academy of Philadelphia in 1749, laying out the principles for the school in a pamphlet titled "Proposals for the Education of Youth of Pennsylvania," which sought to combine a classical Latin grammar education with training in the practical skills needed to make a living. Several years later, Franklin appointed William Smith the school's provost, and within a year, the institution added a degree-granting program called the College of Philadelphia. Amidst the turmoil of the War of Independence, the state of Pennsylvania seized the College of Philadelphia in 1779, transforming it into the University of the State of Pennsylvania. Twelve years later, after the tumult the war caused subsided, state legislators relinquished control of the institution and the privately run University of Pennsylvania was born.
4. Prescott, *Boston Tech*, 5.
5. The College of William and Mary, located in Williamsburg, Virginia, due west of Yorktown between the James and York Rivers on the Virginia Peninsula, is the nation's second oldest college. It was established in 1693 when the reigning joint monarchs of England, William III and Mary II, signed a royal charter for a "perpetual College of Divinity, Philosophy, Languages, and other good Arts and Sciences" in the Virginia Colony (after an ill-fated attempt to establish a similar institution closer to present-day Richmond failed). The site, chosen by its first president, the Rev. James Blair, was on high ground west of the small community of Middle Plantation in James City County. When the colonial capital was moved from Jamestown to Middle Plantation in 1699, the community was renamed Williamsburg in honor of King William III.
6. Heeding Josiah Holbrooke's call to provide mutual education societies—lyceums—for the industrial class, members of Baltimore's merchant and manufacturing elite opened the Maryland Institute for the Promotion of the Mechanic Arts in November of 1825. Located in "The Athenaeum" at the southwest corner of East Lexington and St. Paul Streets, the Institute offered a variety of lecture series on subjects connected with the mechanic arts, exhibitions of articles of American manufacturing, and maintained a library until fire destroyed the building in 1835. After languishing for more than a dozen years, the Institute was reincorporated by the state legislature and endowed with a small annual appropriation and, by 1851, had moved above the old Centre Market on Market Place between East Baltimore and Water Streets, occupying the second floor, where it maintained classrooms, offices, shops and studios, and a large assembly hall until the winter of 1904, when another fire—the Great Baltimore Fire—gutted the building. When the Institute reopened in 1908, its mission had expanded,

offering classes in pottery, metal-working, wood carving, drafting, and textile design. It also maintained several art galleries and exhibition rooms. Today, the Institute, which changed its name in 1959 to the Maryland Institute, College of Art (MICA), has a sprawling campus at the intersection of Mount Royal and West Lafayette Avenues, north of Pearlstone Park in the Bolton Hill district of north Baltimore.
7. Rogers, *Life and Letters*, 1:36-37.
8. Ibid., 1:47.
9. Ibid., 1:49.
10. Ibid., 1:49.
11. The kinship of the Rogers brothers—or, as historians often refer to them, the "Brothers Rogers"—is no exaggeration: the closeness the brothers felt toward each other, probably the result of losing their mother at a young age (coupled with their father's increased affection and guidance), was exhibited throughout their lives both in their ongoing correspondence and in their support of each other's professional needs. For instance, not trusting his assistants, William asked Henry to help him on the Virginia survey, even though Henry was still involved with New Jersey's state survey. When Henry quit work on the Virginia survey to take charge of the Pennsylvania survey, William turned to Robert, who had just received his medical degree from the University of Pennsylvania, asking him to help with the survey instead of starting his medical practice (which he did without complaint). Likewise, when Robert left William to help Henry with the Pennsylvania Survey, James, who had just resigned from the medical school in Cincinnati, stepped in for the departing Robert. Yes, the Brothers Rogers is no exaggeration: throughout their adult lives, they were exceedingly supportive of each other and their professional interests.
12. Rogers, *Life and Letters*, 1:86.
13. Ibid., 1:99.
14. Ibid., 1:96-97.
15. Ibid., 1:96-97.
16. Ibid., 1:123.
17. A. J. Angulo, *William Barton Rogers and the Idea of MIT* (Baltimore: The Johns Hopkins University Press, 2009), 76.
18. Rogers, *Life and Letters*, 1:76.
19. Angulo, *William Barton Rogers*, 177.
20. Patsy Gerstner, *Henry Darwin Rogers, 1808-1866, American Geologist* (Tuscaloosa, AL: The University of Alabama Press, 1994), 105.
21. During William's year as chairman of the University of Virginia's faculty, student riots carried on to such a pitch that the civil authorities had to be summoned in order to suppress them. Before they were summoned, however, Robert, a faculty member at the University since 1842, intercepted a band of student dissenters marching with horns

and drums. Concerned about the disruptive nature of the rioters and the possibility of a melee, Robert hid behind a column in an alcove in front of his house. As the rebel-rousers paraded by with their horns blaring and drums pounding, one of the students approached the front door of Robert's house, whereupon Robert jumped out from behind the column, seized the boy by his cloak, and dragged him into the house. The surprise attack shocked the unruly crowd, bringing an abrupt end to their protest. After the students dispersed, Robert held their comrade only long enough to reprimand him before letting him go.
22. Rogers, *Life and Letters,* 1:257-58.
23. Ibid., 1:258.
24. Ibid., 1:259.
25. Ibid., 1:259.
26. Angulo, *William Barton Rogers,* 81.
27. Prescott, *Boston Tech,* 26.
28. Rogers, *Life and Letters,* 1:422.
29. Ibid., 1:422.
30. Ibid., 1:423.
31. Ibid., 1:426.
32. Ibid., 1:426-27.
33. The original expression coined by Oliver Wendell Holmes was "the Hub of the Solar System" and referred not to Boston, but to the Massachusetts State House perched atop the remains of Beacon Hill. Later, Holmes expanded the phrase to "the Hub of the Universe," a reference to the city of Boston itself, http://www.celebrateboston.com/culture/the-hub-origin.htm.
34. Rogers, *Life and Letters,* 1:264.

CHAPTER 9 | *Objects and Plans*

1. Rogers, *Life and Letters,* 2:19.
2. Ibid., 1:392.
3. Julius Stratton and Loretta Mannix, *Mind and Hand: The Birth of MIT* (Cambridge: The MIT Press, 2005), 144.
4. Ibid., 145.
5. The complete areas listed in the proposal were: (1) agriculture, horticulture, and pomology; (2) natural history, geology, and chemistry; (3) mechanics, manufactures, commerce, and technology; and (4) fine arts and education Source: Prescott, *Boston Tech,* 28.
6. Rogers, *Life and Letters,* 2:4.
7. Stratton and Mannix, *Mind and Hand,* 147.
8. Ibid., 173.
9. Rogers, *Life and Letters,* 2:15.
10. Ibid., 2:21.

11. Ibid., 2:21-22.
12. Ibid., 2:29.
13. Ibid., 2:29.
14. Although I've focused primarily on the efforts of William Rogers in founding MIT, I would be remiss to ignore several other individuals who gave significant time and effort to the cause. Of the eighteen members of the Reservation Committee responsible for creating, promoting, and ultimately seeing MIT's proposal through the state legislature, several names stand out as instrumental to the proposal's successful outcome. *Matthias Ross*, a successful businessman in the cotton industry who oversaw the construction of cotton mills in Lowell, Massachusetts, before founding Ross, Turner & Co. for the production of linen thread used in the manufacturing of fishing lines and nets, was an active force on the Reservation Committee and subsequently a member of MIT's governing board from 1862 to 1892. *George Emerson* was a respected educational leader who taught in private schools until founding his own school for young ladies in 1823. Prior to this, he served briefly as a tutor in mathematics at Harvard and as the English High School's first headmaster. He was also active in founding the Boston Mechanics' Institute and the Boston Society of Natural History, serving as the latter's second president for seven years. It was Emerson, while serving as a member of the Reservation Committee, who proposed that the 1860 memorial emphasize the preparation of a plan for a polytechnic institute. *Erastus Bigelow*, known as one of the greatest inventors of his time, devoted much of his energy to the improvement of the manufacture of fabrics, starting with his first invention—an automatic machine for piping cord—at the age of fourteen. His most important invention, however, was the power loom for weaving carpet that led him to the founding of the Bigelow Carpet Company. A successful businessman and founding member of the National Association of Wool Manufacturers, Bigelow served on MIT's governing board from 1862 until his death in 1879. Rogers regarded Bigelow as one of the "honored band of associates whose active zeal contributed to the foundation of the Institute." Finally, businessman *Charles Dalton* cannot be overlooked. It was his organizational and administrative abilities coupled with his financial acumen sharpened as a partner in the Boston firm of J. C. Howe that lent a steady hand to MIT's financial administration during its early years. Dalton, however, not only served as MIT's first treasurer and member of the governing board, but also had active leadership positions in several Boston-area organizations: he was trustee and president of the Massachusetts General Hospital, vice president of the New England Trust Company, and chairman of the Boston Park Commission. Summarized from Stratton and Mannix, *Mind and Hand*, 163-183; Rogers' quotation above, *Life and Letters*, 2:171.

15. Prescott, *Boston Tech*, 29
16. Although the original intent of the Associated Institutions' charter was to make the three units or branches of the Institute co-equal, each one represented by a committee of government with its own officers and members, over time, the term "Institute of Technology" became associated almost exclusively with the School of Industrial Science. Part of the confusion came from Rogers' role as both president of the Institute and principal of the School, but also because over time the School overshadowed the other two branches of the Institute. The confusion was more than rhetorical; it was legal. As such, in 1869, the institute's charter was amended to make the Society of Arts a distinct organization subsidiary to the Institute (the Museum of Arts never formally materialized), with its own officers, constitution, and bylaws. The same amending act changed the name of the Institute's governing body from "Government" (which the Society of Arts still utilized) to the more comprehensive label "Corporation," a term most of MIT's early supporters could relate to since many of them had affiliations with Harvard, which was governed by "The Corporation" comprised of the President and the Fellows of Harvard. Source: Prescott, *Boston Tech*, 36-38.
17. Rogers, *Life and Letters*, 2:41.
18. Ibid., 2:45.
19. Stratton and Mannix, *Mind and Hand*, 198.
20. Ibid., 201.
21. Rogers, *Life and Letters*, 2:4.
22. Stratton and Minnix, *Mind and Hand*, 207.
23. Prescott, *Boston Tech*, 31.
24. Stratton and Minnix, *Mind and Hand*, 211.
25. Ibid., 215.
26. Prescott, *Boston Tech*, 32.
27. Stratton and Minnix, *Mind and Hand*, 220.
28. Roger Geiger, *The History of American Higher Education: Learning and Culture from the Founding to World War II* (Princeton, NJ: Princeton University Press, 2015), 281.
29. Angulo, *William Barton Rogers*, 111.
30. Geiger, *History of American Higher Education*, 281.
31. Ibid., 306.
32. Angulo, *William Barton Rogers*, 113.
33. In one sense, Agassiz and Rogers were a perfect match for each other: intelligent, ambitious, verbally adept, charismatic, and highly regarded by their peers in and out of the scientific community. Their manifold talents were put on full display in a series of debates held in the early months of 1860 after the release of Darwin's groundbreaking work, *On the Origin of the Species by Means of Natural Selection*, in November

1859. Sponsored by the American Academy of Arts and Sciences and held at the Boston Society of Natural History, the debates drew an enthusiastic audience of scientists, civic leaders, and academics. This was not the first time Darwin's ideas had been debated, in America or abroad. Agassiz and Harvard botanist Asa Gray, who had corresponded directly with Darwin about his revolutionary ideas prior to the release of Darwin's book, held similar debates only a year or so earlier, with Agassiz dominating the debates with his more thorough knowledge of the scientific data, primarily of a geological nature, and his persuasive rhetorical skills. With Rogers, the outcome would be different: whereas Agassiz and Rogers' rhetorical skills were equally matched, Rogers held the advantage in the scientific data given his experience directing the Virginia State Geological Survey. This, coupled with Rogers' powers of logic, was enough to put Agassiz on the defensive quite early. The debates hinged on Agassiz's refutation of Darwin's theories. Unlike Darwin (and Rogers), Agassiz believed in "special creation," that species originated from an idea of the Creator; that they were fixed, immutable, and formed a great chain or continuum from lowest to highest, with humans the most superior and intelligent of all creative acts. Upon reading Darwin's work, Agassiz had an almost visceral reaction to it, calling it a monstrosity, a "fanciful theory" (Angulo, p. 53). But for every argument Agassiz presented in renunciation of Darwin's theories, Rogers was ready with evidence-based counter arguments that, by the sixth debate, put Agassiz well back on his heels. Like a boxer against the ropes, Agassiz stumbled, confusing technical terms, misquoting the scientific literature, and just plain out contradicting himself. Rogers was quick to point out Agassiz's weaknesses, deftly turning Agassiz's liabilities into advantages. By all accounts, even supporters of Agassiz, Rogers emerged victorious from the debates, but it cost him professionally, since Agassiz had tremendous clout. Although Rogers was on several occasions considered for a position at the Lawrence Scientific School, no offer was ever forthcoming since Agassiz would have had to approve it, and after the debates, there was little chance that Agassiz would have done that.
34. Angulo, *William Barton Rogers,* 115.

CHAPTER 10 | *The New Education*

1. Stratton and Mannix, *Mind and Hand,* 478.
2. Of the thirteen committees, the first to form after the Act of Incorporation was signed into law were: the Committee on Publication (John Runkle, chairman); the Committee on the Museum (Erastus Bigelow, chairman); the Committee on Instruction (William Rogers, chairman); and the Committee on Finance (Matthias Ross, chairman). It is worth

noting the original officers of MIT as well: William Rogers, president; Jacob Bigelow, John Chase, John Amory Lowell, and Marshall Wilder, vice presidents; Charles Dalton, treasurer; and Thomas Webb, secretary. For the record, Rogers was initially elected "principal" of MIT, not president, a title more in line with headmaster of a high school than president of a college. Several months later, on November 14, 1865, Rogers' title was changed to president, though he remained the head or "principal" of the School of Industrial Science as well.
3. The full title of the petition being: *Objects and Plan of an Institute of Technology; Including a Society of Arts, a Museum of Arts, and a School of Industrial Science. Proposed to be Established in Boston. Prepared by Direction of the Committee of Associated Institutions of Science and Arts; and Addressed to Manufacturers, Merchants, Mechanics, Agriculturalists, and Other Friends of Enlightened Industry in the Commonwealth* (Boston: John Wilson and Son, 1861), 1-29.
4. Stratton and Mannix, *Mind and Hand*, 427.
5. Ibid., 438.
6. Ibid., 437.
7. Of all the institutions listed, Rensselaer Polytechnic Institute, under the direction of Benjamin Franklin Greene (who succeeded Amos Eaton, the Institute's first director), was the most like the Massachusetts Institute of Technology in its early days, offering by MIT's inaugural year, four major programs, three in engineering (civil, mechanical, and topographical, with a fourth—mining engineering—proposed) and one in natural science. No other institution of higher learning, except perhaps Yale's Sheffield Scientific School, could boast such an emphasis on engineering, not even West Point, and certainly not Harvard's Lawrence Scientific School.
8. According to Stratton and Mannix (445), when the first college catalogue was printed in December 1865, the dual-department approach was shelved, and in its place three strategic goals outlined the Institute's mission: "*First*, To provide a full course of scientific studies and practical exercises for students seeking to qualify themselves for the professions of the Mechanical Engineer, Civil Engineer, Practical Chemist, Engineer of Mines, and Builder and Architect. *Second*, To furnish such a general education, founded upon the Mathematical, Physical, and Natural Sciences, English and other Modern Languages, and Mental and Political Science, as shall form a fitting preparation for any of the departments of active life. *Third*, To provide courses of Evening Instruction in the main branches of knowledge above referred to, for persons of either sex who are prevented, by occupation and other causes, from devoting themselves to scientific study during the day, but who desire to avail themselves of systematic evening lessons or lectures."

9. Rogers' *Scope and Plan* listed six departments, the last being General Science and Literature, which was overseen by William Atkinson, Professor of English Language and Literature. As the first course catalogue suggested, the intent of the program—later shortened to Science and Literature—was the preparation of students "for any of the departments of active life" (*First Annual Catalogue*, 9). In other words, it was for the student who did not want to specialize in any one area of science, but wanted more science classes than the traditional liberal arts curriculum afforded. As Stratton and Mannix (548) note, the Sheffield Scientific School at Yale had a similar program designed for students who "wished a general knowledge of science, without specializing in any particular field."
10. Stratton and Mannix, *Mind and Hand*, 449.
11. It is almost taken for granted that the doors of nineteenth-century higher education were not yet open to women, especially at colleges established during the colonial and early antebellum period. At MIT, women were welcome, but only in the evening classes, which were open to any adult seeking "such useful knowledge as they can acquire without methodical study and in hours not occupied by business" (Stratton and Mannix, 427). Although full-time female students would be admitted to MIT, it would take a few years before the institution was not only willing to admit them, but also suitably adapted to admit them (the issue in question: co-educational laboratories). Although the first serious inquiries by women came as early as 1867, it was not until 1870 that the first female student was finally accepted into a full-time professional program. Ellen Swallow, a Vassar College graduate, was accepted as a special student in the chemical laboratory in the fall of 1870. Three years later, Ms. Swallow received the Institute's first degree awarded to a woman.
12. The design and construction of MIT's first building in the Back Bay on Boylston Street was a long, arduous, and costly experience. Ultimately, the Building Committee would turn to the father-son architectural firm of Jonathan and William Preston, with the younger Preston designing three of the buildings to occupy the site in the Back Bay allotted to the Associated Institutions. The Boston Society of Natural History's building, completed in 1862, faced Berkeley Street, while the Institute of Technology's massive five-story building, completed several years later, faced Boylston Street. A third building, for the Institute's museum, was never built. William Preston, who studied architecture at the École des Beaux-Arts in Paris after one year at the Lawrence Scientific School, was only nineteen years old when the Boston Society of Natural History chose his design for their Back Bay building. It was his first commission in an impressive career that would span almost half a century. An early historic preservationist, Preston was instrumental in the successful 1896

effort to prevent the Massachusetts General Court from demolishing Boston's historic State House designed by Boston architect Charles Bulfinch in 1798.
13. The first classes offered in the Mercantile Building on Summer Street in the winter of 1865 were: Elementary Mathematics, with Practice in the Use of the Chain, Level, etc.; Elementary Physics; Elementary Chemistry, with Manipulations; Drawing; and the French language. A notice to this extent, signed by William B. Rogers, President, Massachusetts Institute of Technology, prior to the February 1865 start date, was circulated among prospective students and included the following: "The preliminary course will cover a period of four months, commencing, it is expected, about the middle of February. The precise date of its commencement, as well as the programme of instruction in the several classes, will be made known as soon as their organization has been determined on" (Prescott, *Boston Tech*, 44).
14. Stratton and Mannix, *Mind and Hand*, 323.
15. One example of a faculty member often wearing many hats is William Atkinson, professor of English language and literature. Originally hired to teach English composition, rhetoric, and literature, Atkinson would soon add history and political economy to his teaching responsibilities, prompting a change in title in 1871 to Professor of English and History. In addition to his teaching load, Atkinson oversaw the Science and Literature Course, served as faculty secretary, and filled in as the school's librarian, without salary, until the position was filled several years later.
16. James Hague, who studied at the Lawrence Scientific School and at several prestigious universities abroad, was listed for several years as a faculty member in MIT's annual course catalogues—but he never taught one course. His name was finally dropped for the 1868-69 catalogue, replaced by Alfred P. Rockwell, professor of mining engineering at Yale's Sheffield Scientific School.
17. Bruce Sinclair, "Harvard, MIT, and the Ideal Technical Education," in *Science at Harvard University: Historical Perspectives*, edited by Clark A. Elliott and Margaret W. Rossiter (Bethlehem, PA: Lehigh University Press, 1992), 78.
18. Hugh Hawkins, *Between Harvard and America: The Educational Leadership of Charles W. Eliot* (New York: Oxford University Press, 1972), 3.
19. Ibid., 5.
20. Ibid., 11.
21. One of Eliot and Storer's first collaborations was developing a new course in chemistry to be offered as an elective during the junior year, making it one of the first courses at Harvard to extend student elective options beyond languages and mathematics. It was also the first course

in the academic college to place laboratory exercises at the center of the student's experience.
22. As a consulting chemist, Strorer had many clients in the Boston area, one of them the Boston Gas Light Company, for which he tested the quality of gas furnished to its customers. The position put Storer in contact with William Rogers, who had been appointed the Commonwealth's gas inspector prior to establishing the Institute of Technology. Several years later, it was Rogers who enlisted Storer to teach general and analytical chemistry at the new school. Storer jumped at the chance to work with Rogers, eager to teach chemistry by the laboratory method, a method strongly favored by Rogers, not only for chemistry, but also for all the sciences.
23. Hawkins, *Between Harvard and America*, 14.
24. Ibid., 16.
25. Ibid., 16.
26. Ibid., 16.
27. Ibid., 20.
28. If Eliot's plan seems strangely similar to the plan put forth by William Rogers in chartering MIT, consider that Eliot and Rogers were no strangers to each other. Not only did they both have ties to Harvard and the Lowell Institute, but also they were both members of the Thursday Evening Club, a private men's club founded in Boston in 1846 to discuss advances in the arts and sciences. In fact, Rogers was present the evening that Eliot shared his plan with members of the club. It would not be the first interaction of the two reformers, nor would it be the last.
29. The Lazzaroni, which translates "beggar" in Italian, a playful literary reference to a group of Italian beggars associated with the hospital of St. Lazarus in Naples, was made up of an elite group of American scientists, all leaders in their respective fields, who saw themselves as "begging" for money to support their interests in science, since they received little or no financial support from either their university or the government, depending solely on student fees and tuition and the munificence of wealthy industrialists. The group, which first went by the name the Florentine Academy, consisted of some of the greatest American scientists of the day: Alexander Dallas Bache, Benjamin Peirce, Joseph Henry, Louis Agassiz, Benjamin Gould, Wolcott Gibbs, and James Dwight Dana. The Lazzaroni backed the idea of a "national university," where they could bring together faculty interested in pure or theoretical science in a center for scientific research and teaching. The university would feature advanced lectures by the nation's most creative and gifted scientists and scholars. The following quote from Joseph Henry aptly sums up the Lazzaroni's position regarding the state of science in America in the middle of the nineteenth century: "We are

overrun in this country with charlatanism. Our newspapers are filled with puffs of Quackery and every man who can burn phosphorus in oxygen and exhibit a few experiments to a class of young ladies is called a man of science" (Resetarits, 82). Along with establishing a national university, their goal was to create a national standard for judging what is good science. Influenced by the German university model, they preferred pure or theoretical research over and above applied science conducted by so-called "experimentalists." The Lazzaroni's emphasis on the value of pure science is summed up by Theodore Lyman: "The moment a chemist tells me that such and such facts apply to the manufacture of a thing, there I say good-bye, for of practical things I have more than enough in my daily life" (Sinclair, 79). Active on the national scene, the Lazzaroni influenced the founding of the American Association for the Advancement of Science and the National Academy of Sciences. Ultimately, internal squabbles and the death of two of its leaders, Bache and Henry, led to the dissolution of the Lazzaroni. However, before it dissolved, their last triumph was the appointment of Wolcott Gibbs as the fourth Rumford Professor of Applied Science at Harvard's Lawrence Scientific School.

30. F. W. Clarke, "Biographical Memoir of Wolcott Gibbs, 1822-1908," *National Academy of Sciences Biographical Memoirs* 7 (February 1910): 3.

31. It's not that Harvard didn't try to retain Eliot; after all, Eliot was a valued member of the institution. It's just that in 1863, with the country ravaged by two years of civil war, members of Harvard Corporation were in no mood to make long-term decisions for the college. With the Rumford professorship going to Gibbs, the Corporation offered Eliot all it could—an upgrade from assistant professor to professor. There was a catch, however: the upgrade was only in title; Eliot's salary would remain the same (with any increase dependent upon an increase in student enrollment). According to Hawkins (26), "To lose the chance to make Storer his colleague was a disappointment, to be passed over for the Rumford chair a humiliation, but to be offered an under-salaried professorship where the exact income depended on his ability to draw students was an outrage." When the Corporation suggested various schemes to raise additional money for Eliot's salary, including soliciting money from Eliot's relatives, the indignant Eliot resigned.

32. As mentioned in the previous footnote, the chance to work with Storer again also made the Rumford professorship appealing to Eliot. Eliot imagined that he could persuade President Hill and the Corporation to create a professorship in chemistry for Storer so the two of them—working collaboratively—could build the chemistry program at the scientific school, a program that had been sorely neglected by Eben Horsford.

33. Hawkins, *Between Harvard and America*, 30.
34. Ibid., 36.
35. The authors are listed as Charles W. Eliot and Frank H. Storer. The complete title reads: *A Manual of Inorganic Chemistry: Arranged to Facilitate the Experimental Demonstration of the Facts and Principles of the Science.* The textbook was first published in 1867, with the note, "Printed for the Authors," and lists the Boston firm of Rockwell & Rollins as the printers. A second edition released the following year lists the New York firm of Ivison, Phinney, Blakeman, and Co. as the book's publisher. The dedication in the two editions is the same, and noteworthy: "The authors inscribe this book to their teacher of chemistry, Prof. Josiah P. Cooke, of Harvard College, in token of gratitude and friendship."
36. Paul Croce, *Science and Religion in the Era of William James, Volume I: Eclipse of Certainty, 1820*-1880 (Chapel Hill & London: The University of North Carolina Press, 1995), 136.
37. Eliot and Storer, *Manual of Inorganic Chemistry*, iii.
38. *Ibid.*, iii.
39. Hawkins, *Between Harvard and America*, 41.
40. Ibid., 42.
41. Charles Eliot, "The New Education," *The Atlantic Monthly* 23 (February 1869): 203.
42. Ibid., 204.
43. Ibid., 206.
44. Ibid., 210.
45. Ibid., 210.
46. Ibid., 214.
47. Lawrence Aronovitch, "The Spirit of Investigation: Physics at Harvard University, 1870-1910." In *The Development of the Laboratory: Essays on the Place of Experiment in Industrial Civilization*, edited by Frank A. J. L. James (London: Macmillan Press, 1989), 85.
48. Ibid., 86.
49. Hawkins, *Between Harvard and America*, 43.
50. Bruce Sinclair, "Harvard," 79.
51. Hawkins, *Between Harvard and America*, 45.
52. Ibid., 46.
53. Eliot, "The New Education," 203.

CHAPTER 11 | *With His Own Hands*

1. Edward Pickering, "Physical Laboratories," *Nature* 3 (January 26, 1871): 241.
2. Prescott, *Boston Tech*, 54-64.
3. Stratton and Mannix, *Hand and Mind*, 581.

4. John Daniel Runkle, affectionately called "Uncle Johnny" by his students at the Institute of Technology, was a member of the first graduating class of the Lawrence Scientific School (Class of 1851). He received an S.B. (Bachelor of Science) and, for his high scholarship, an honorary A.M. (Master of Arts). Runkle was one of the earliest supporters of Rogers' idea for a "School of Industrial Science" (which soon became known as the Massachusetts Institute of Technology). Runkle, who taught mathematics, became acting president of the Institute in the fall of 1868, after Rogers was incapacitated by illness. He would serve as such for the next two years, until being named president of the Institute after Rogers' official resignation. Runkle would serve as president for eight years, being succeeded briefly by Rogers until the appointment of Francis Amasa Walker, MIT's third president, who began his term in office in the fall of 1881.
5. Nathaniel Thayer? Now that name rings a bell. Yes, remember on our tour of Beacon Hill's South Slope neighborhood, we stopped in front of a large Italianate stone building next door to Edward Pickering's childhood home on Mount Vernon Street. The structure, a duplex, was built for the Thayer brothers, John and Nathaniel, who made a fortune in banking and railroad speculation. And, yes, it's the same Nathaniel Thayer who gave the Institute of Technology $25,000 to endow a chair of physics, the Thayer Professor of Physics, which Edward Pickering assumed in 1868. Of course, we've heard this all before: it's a small world. Yes, it is—or was—among the socially elite and well-heeled families of nineteenth-century Boston.
6. Cajori, *Laboratories*, 298.
7. Ibid., 300.
8. Ibid., 300-301.
9. Edward Pickering, "Physical Laboratories," 241.
10. Stratton and Mannix, *Hand and Mind*, 584-585.
11. Ibid., 585.
12. Ibid., 585.
13. Edward Pickering, "Department of Physics, Jan. 31, 1877," *Massachusetts Institute of Technology President's Report for the Year Ending Sept. 30, 1876* (Boston: A. A. Kingman, 1877), 34.
14. "Rogers Laboratory of Physics," MIT Dept. of Archives & Special Collections, https://libraries.mit.edu/archives/exhibits/pickering/
15. In 1870, Pickering devised an apparatus for the electrical transmission of sound, which he demonstrated to his classes and later described at a meeting of the American Association for the Advancement of Science in Troy, New York. When approached by several businessmen interested in the commercial possibilities of his "speaking tube," Pickering placed at their disposal the apparatus and its design, even allowing the group to use the Institute of Technology's laboratory for their own investi-

gations. It was a precept that Pickering whole-heartedly supported; according to the young investigator, science should have no secrets, a sentiment captured later when Pickering stated: "Science is an ennobling pursuit only when it is wholly unselfish." Source: Annie Cannon, "Edward Charles Pickering," *Popular Astronomy* 27, no. 3 (March 1919): 181.
16. Edward Pickering, *Elements of Physical Manipulation* (New York: Hurd & Houghton, 1873, 1876; Boston: Houghton, Mifflin & Co., 1895).
17. Pickering, *Elements,* dedication page.
18. Ibid., v.
19. Ibid., vi.
20. Ibid., vi.
21. Ibid., vi.
22. Ibid., vi-vii.
23. Stratton and Mannix, *Hand and Mind,* 586.
24. Ibid., 587.
25. Ibid., 588.
26. Edward Pickering, "Department of Physics, Jan. 31, 1877," *Massachusetts Institute of Technology President's Report for the Year Ending Sept. 30, 1876* (Boston: A. A. Kingman, 1877), 31-32.
27. Edward Pickering would be known less for his work at MIT, and more for his forty-two-year reign as director of the Harvard Observatory where, along with a cadre of young men and women, he worked to modernize the field of astronomy through the use of photography and spectrometry. Not only was he an able administrator and fundraiser, but he was also a farsighted scientist who understood the impact of new technologies on the field of astronomy. For his work, Pickering was awarded the Gold Medal of the Royal Astronomical Society in 1886 and 1901, the Valz Prize of the French Academy of Sciences and the Henry Draper Medal from the National Academy of Sciences in 1888, the Bruce Medal in 1908, and the Prix Jules Janssen Award from the French Astronomical Society in 1908. Regarding Edward Pickering's mentor Charles Eliot, Harvard's longest-serving president would bring much-needed and lasting reforms to the institution that he dearly loved, elevating it from a mid-nineteenth-century provincial college to a leading twentieth-century university with a full array of elective courses for undergraduates and a wide spectrum of professional programs for advanced students. Although Eliot pushed through a number of reforms during his forty-year presidency, he failed to complete an early initiative: a Harvard-MIT merger. Although the idea was not his originally, Eliot embraced the idea of consolidating the science and technology interests of the two institutions, bringing them under one roof—Harvard's, of course. But William Rogers and John Runkle, as well as most of MIT's faculty, vehemently opposed the idea, fearful of

giving up their autonomy to the larger and more traditionally oriented institution in Cambridge. The one faculty member who supported Eliot's call for a merger was Eliot's friend and colleague Frank Storer. So inflamed was Storer by Rogers and Runkle's refusal to consider the merger that the young chemistry professor quit abruptly in 1870, surfacing shortly thereafter as dean of the Bussey Institution, Harvard's undergraduate school for the study of agricultural science. Another casualty of the failure to merge the two institutions was the Lawrence Scientific School, which withered on the vine after Eliot began his push to establish a graduate studies program, consolidating all graduate programs in one school by the end of the century.

28. Annie Cannon, "Edward Charles Pickering," 177.

Bibliography

"Abbott Lawrence." *The American Journal of Education and College Review*, 2 (January 1856): 205-216.

Adams, Henry. *The Education of Henry Adams*. New York: Library of America Paperback Classics, 2010.

Allison, Robert J. *Short History of Boston*. Carlisle, MA: Commonwealth Editions, 2004.

American Lyceum, with the Proceedings of the Convention Held in New York, May 4, 1831, to Organize the National Department of the Institution. Boston: Hiram Tupper, Printer, 1831.

An Oration by Phillips Brooks, D.D. and a Poem by Robert Grant at the Celebration of the Two Hundred and Fiftieth Anniversary of the Foundation of the Boston Latin School, April 23, 1885. Boston: Houghton, Mifflin & Co., 1885.

Andrew, John. *A View of Beacon Street, Boston*. Boston Public Library's Pictorial Archive, http://www.digitalcommonwealth.org/search/commonwealth:4q77fw147.

Angulo, A. J. *William Barton Rogers and the Idea of MIT*. Baltimore, MD: Johns Hopkins University Press, 2009.

Army, Jr., Thomas F. *Engineering Victory: How Technology Won the Civil War*. Baltimore, MD: Johns Hopkins University Press, 2016.

Aronovitch, Lawrence. "The Spirit of Investigation: Physics at Harvard University, 1870-1910." In *The Development of the Laboratory: Essays on the Place of Experiment in Industrial Civilization*, edited by Frank A. J. L. James. London: Macmillan, 1989.

Bailey, Solon I. "Biographical Memoir of Edward Charles Pickering, 1846-1919." *National Academy of Sciences Biographical Memoirs*, 15 (1932): 167-189.

Baldwin, Ebenezer. *Annals of Yale College*. New Haven, CT: Hezekiah Howe, 1831.

Ballard, Frank W. *The Stewardship of Wealth, as Illustrated in the Lives of Amos and Abbott Lawrence: A Lecture Before the New York Young Men's Christian Association, January 4, 1865*. New York: Clayton & Medole, 1865.

Beckham, Stephen Dow, "Colonel George Gibbs." In *Benjamin Silliman and His Circle: Studies on the Influence of Benjamin Silliman on Science in America*, edited by Leonard G. Wilson. New York: Science History Publications, 1979.

Betros, Lance, ed. *West Point; Two Centuries and Beyond*. Abilene, TX: McWhitney Foundation Press, 2004.

Bode, Carl. *The American Lyceum: Town Meeting of the Mind*. Carbondale, IL: Southern Illinois University Press, 1968.

Boston Directory, for the Year 1852, Embracing the City Record, A General Directory of the Citizens, and a Business Directory, with an Almanac, from July, 1852, to July, 1853. Boston: George Adams, 1852.

Brayley, Arthur Wellington. *Schools and Schoolboys of Old Boston*. Boston: Louis P. Hager, 1894.

Brown, Chandos Michael. *Benjamin Silliman: A Life in the Young Republic*. Princeton, NJ: Princeton University Press, 1989.

Bunting, Bainbridge. *Harvard: An Architectural History*. Completed and Edited by Margaret Henderson Floyd. Cambridge, MA: Belknap Press of Harvard University, 1985.

_____. *Houses of Boston's Back Bay: An Architectural History, 1840-1917*. Cambridge, MA: Belknap Press of Harvard University, 1967.

Burke, Sarah K. "Bookish Fires: The Legacy of Fire in the Harvard Libraries." *Library Preservation at Harvard* (Spring 2010): http://preserve.harvard.edu.

Cajori, Florian. *A History of Physics in its Elementary Branches, Including the Evolution of Physical Laboratories*. New York: MacMillan, 1922.

Campion, Nardi Reeder. "Who Was Sylvanus Thayer?" *Dartmouth Engineer Magazine* (Fall 2004), https://engineering.dartmouth.edu/magazine/who-was-sylvanus-thayer.

Cannon, Annie J. "Edward Charles Pickering," *Popular Astronomy: A Review of Astronomy and Allied Sciences*, 27, no. 3 (March 1919): 177-183.

Carpenter, James. "Thomas Jefferson and the Ideology of Democratic Schooling." *Democracy & Education*, 21, no. 2 (2013): 1-11.

Carr, Jacqueline Barbara. *After the Siege: A Social History of Boston 1775-1800*. Boston: Northeastern University Press, 2005.

Carver, Robin. *History of Boston*. Boston: Lilly, Wait, Colman, & Holden, 1834.

Catalogue of Graduates of the Lawrence Scientific School of Harvard University, 1851-1895. Cambridge, MA: Harvard University Press, 1895.

Catalogue of the Boston Latin School, Established in 1635, with an Historical Sketch prepared by Henry F. Jenks. Boston: Boston Latin School Association, 1886.

Catalogue of the Officers and Students of Harvard University for the Academical Year 1864-1865, First Term. Cambridge, MA: Harvard University Press, 1864.

The Centennial Anniversary of the Birth of Edward Everett, Celebrated by the Dorchester Historical Society, April 11th, 1894. Boston: Municipal Printing Office, 1897.

Chamberlain, Allen. *Beacon Hill: Its Ancient Pastures and Early Mansions.* Boston: Houghton, Mifflin & Co., 1925.

Chittenden, Russell H. *History of the Sheffield Scientific School of Yale University, 1846-1922, Vol. II.* New Haven, CT: Yale University Press, 1928.

Clarke, F. W. "Biographical Memoir of Wolcott Gibbs, 1822-1908." *National Academy of Sciences Biographical Memoirs* 7 (February 1910): 1-22.

Claxton, Timothy. *Memoir of a Mechanic: Being a Sketch of the Life of Timothy Claxton, Written by Himself, Together With Miscellaneous Papers.* Boston: George W. Light, 1839.

Constitution of the Massachusetts Charitable Mechanic Association, Instituted March 15, 1795, and Incorporated March 8, 1806. Boston: Press of Crocker & Brewster, 1855.

Cordasco, Francesco. *The Shaping of American Graduate Education: Daniel Coit Gilman and the Protean Ph.D.* Totowa, NJ: Rowman & Littlefield, 1973.

Cramer, Kenneth C. "George Ticknor, 1807." *Notes from the Special Collections, Dartmouth College Library Bulletin* (November 1991): https://www.dartmouth.edu/~library/Library_Bulletin/Nov1991/LB-N91-KCramer.html.

Croce, Paul Jerome. *Science and Religion in the Era of William James, Volume I: Eclipse of Certainty, 1820-1880.* Chapel Hill & London: University of North Carolina Press, 1995.

Crowley, Sharon. *Composition in the University: Historical and Polemical Essays.* Pittsburgh, PA: University of Pittsburgh Press, 1998.

Cushing, Thomas. *Historical Sketch of Chauncy-Hall School with Catalogue of Teachers and Pupils, and Appendix, 1826-1894.* Boston: David Clapp & Son, 1895.

Dana, Jr., Richard H. *An Address Upon the Life and Services of Edward Everett: Delivered Before the Municipal Authorities and Citizens of Cambridge, February 22 1865.* Boston: Sever & Francis, 1865.

Dartmouth College: The Relation Between the College and the Schools. Hanover, NH: Dartmouth College, 1893.

Davis, Ted. "The Dean of American Science." *BioLogos* (August 27, 2015): https://biologos.org/blogs/ted-davis-reading-the-book-of-nature/the-dean-of-american-science.

Deese, Helen R., and Guy R. Woodall. "A Calendar of Lectures Presented by the Boston Society for the Diffusion of Useful Knowledge, 1829-1847." *Studies in the American Renaissance* (1986): 17-67.

Denham, Thomas J. "A Historical Review of Curriculum in American Higher Education: 1636-1900." A course paper presented in partial fulfillment of the requirements for the degree of Doctor of Education. Massachusetts Cluster, Nova Southeastern University (July 2002) ERIC DOC 471739.

Diamant, Lincoln. *Chaining the Hudson, the Fight for the River in the American Revolution*. New York: Carol Publishing Group, 1994.

Donahue, Brian. "Remaking Boston, Remaking Massachusetts." In *Remaking Boston: An Environmental History of the City and Its Surroundings*, edited by Anthony N. Penna and Conrad Edick Wright. Pittsburgh, PA: University of Pittsburgh Press, 2009.

Dorrien, Gary. *The Making of Liberal Theology: Imagining Progressive Religion, 1805-1900*. Louisville, KY: Westminster John Knox Press, 2001.

Drake, William E. *The American School in Transition*. New York: Prentice-Hall, 1955.

Dupuy, R. Ernest. *Sylvanus Thayer: Father of Technology in the United States*. West Point, New York: The Association of Graduates, United States Military Academy, 1958.

Ede, Andrew and Lesley B. Cormack. *A History of Science in Society: From the Scientific Revolution to the Present, vol. 2*. North York, Ontario, Canada: University of Toronto Press, 2012.

Eliot, George Fielding. *Sylvanus Thayer of West Point*. New York: Julian Messner, 1959.

Eliot, Charles William. "The New Education." *The Atlantic Monthly* 23 (February 1869): 203-220; (March 1869): 358-367.

Ellery, Harrison and Charles Pickering Bowditch. *The Pickering Genealogy: Being an Account of the First Three Generations of The Pickering Family of Salem, Mass, and of the Descendants of John and Sarah (Burrill) Pickering, of the Third Generation, Vol. 1-III*. Boston: Privately Printed, 1897.

Elliott, Clark A. "Founding of the Lawrence Scientific School at Harvard University, 1846-1847: A Study in Writing and History." *Archivaria* 38 (1994): 119-130.

———, and Margaret W. Rossiter. *Science at Harvard University: Historical Perspectives*. Bethlehem, PA: Lehigh University Press, 1992.

Enfield, William. *Institutes of Natural Philosophy, Theoretical and Experimental*. London: J. Johnson, 1775.

Everett, Edward. "Inaugural Address." *Addresses at the Inauguration of the Hon. Edward Everett, LL.D., as President of the University at Cambridge, Thursday, April 30, 1846*. Boston: Charles C. Little & James Brown, 1846.

———. *A Memoir of Mr. John Lowell, Jr.* Boston: Charles C. Little & James Brown, 1840.

Farrell, Betty G. *Elite Families: Class and Power in Nineteenth-Century Boston.* Albany, NY: State University of New York Press, 1993.
"The Fire at Harvard College." *The Magazine of History with Notes and Queries* 175 (1931): 173-78.
First Annual Catalogue of the Officers and Students and Programme of the Course of Instruction, of the School of the Massachusetts Institute of Technology, 1865-6. Boston: John Wilson & Sons, 1865.
Ford, Paul Leicester, ed. *The Writings of Thomas Jefferson, Vol. X, 1816-1826.* New York: G. P. Putnam's Sons, 1899.
Franklin, Fabian. *The Life of Daniel Coit Gilman.* New York: Dodd, Mead & Co., 1910.
Fulton, John F. "Benjamin Silliman in His Lighter Moments." In *The Centennial of the Sheffield Scientific School,* edited by George R. Baitsell. New Haven, CT: Yale University Press, 1950.
———, and Elizabeth H. Thomson. *Benjamin Silliman: Pathfinder in American Science.* New York: Henry Schuman, 1947.
Geiger, Roger L. *The History of American Higher Education: Learning and Culture from the Founding to World War II.* Princeton, NJ: Princeton University Press, 2015.
Gerstner, Patsy. *Henry Darwin Rogers, 1808-1866, American Geologist.* History of American Science and Technology Series. Tuscaloosa, AL: University of Alabama Press, 1994.
Gilman, Daniel Coit. "Our National Schools of Science." *North American Review* (October 1867): 498-499.
———. *The Sheffield Scientific School of Yale University: A Semi-Centennial Historical Discourse.* New Haven, CT: Sheffield Scientific School, 1897.
Grayson, Lawrence P. *The Making of an Engineer: An Illustrated History of Engineering Education in the United States and Canada.* New York: John Wiley & Sons, 1993.
Greene, John C. "Protestantism, Science, and American Enterprise: Benjamin Silliman's Moral Universe." In *Benjamin Silliman and His Circle: Studies on the Influence of Benjamin Silliman on Science in America,* edited by Leonard G. Wilson. New York: Science History Publications, 1979.
Greenstein, George. *Portraits of Discovery: Profiles of Scientific Genius.* New York: John Wiley & Sons, 1998.
Guarino, Robert E. *Beacon Street: Its Buildings and Residents.* Charleston, SC: History Press, 2011.
Guralnick, Stanley M. *Science and the Ante-Bellum American College.* Philadelphia, PA: American Philosophical Society, 1975.
Hague, Arnold. "Biographical Memoir of Samuel Franklin Emmons, 1841-1911." *National Academy of Sciences Biographical Memoirs* 7 (1912): 307-334.

Hall, Edwin H. "Biographical Memoir of John Trowbridge, 1843-1923." *National Academy of Sciences Biographical Memoirs* 14 (1930): 183-204.

Harrington, Hugh T. "The Great West Point Chain." *Journal of the American Revolution*, (September 25, 2014): http://allthingsliberty.com/2014/09/the-great-west-point-chain.

Hawkins, Hugh. *Between Harvard and America: The Educational Leadership of Charles W. Eliot*. New York: Oxford University Press, 1972.

Hayward, John. *The New England Gazetteer; Containing Descriptions of all the States, Counties and Towns in New England: Also Descriptions of the Principal Mountains, Rivers, Lakes, Capes, Bays, Harbors, Islands, and Fashionable Resorts within that Territory Alphabetically Arranged*. Concord, NH: William White, 1839.

Hill, Hamilton Andrews. *Memoir of Abbott Lawrence*. Boston: Little, Brown & Co., 1884.

History of Boston from 1630 to 1856. Boston: F. C. Moore & Co., 1856.

Hogarth, Paul. *Walking Tours of Old Boston*. New York: A Brandywine Press Book, E. P. Dutton, 1978.

Holbrook, Josiah. *American Lyceum, or Society for the Improvement of Schools, and the Diffusion of Useful Knowledge*. Boston: T. R. Marvin, 1829.

Holmes, Pauline. *A Tercentenary History of the Boston Latin School, 1635-1935*. Cambridge, MA: Harvard University Press, 1935.

Homans, I. Smith. *History of Boston, from 1630 to 1856*. Boston: F. C. Moore & Co., 1856.

Homer, Rachel Johnston. *The Legacy of Josiah Johnson Hawes: 19th Century Photographs of Boston*. Barre, MA: Barre Publishers, 1972.

Hornberger, Theodore. *Scientific Thought in the American Colleges, 1638-1800*. New York: Octagon Books, 1968.

Horsford, Eben Norton. *The Indian Names of Boston, and Their Meaning*. Cambridge: John Wiley & Son, 1886.

Howe, Marc Anthony de Wolfe. *Boston Common: Scenes from Four Centuries*. Cambridge, MA: Riverside Press, 1910.

Images of America: Boston Common. Published by Friends of the Public Garden. Charleston, SC: Arcadia Publishing, 2005.

The Influence and History of the Boston Athenaeum from 1807 to 1907. Boston: Boston Athenaeum, 1907.

Ireland, Corydon. "Harvard's Long-ago Student Uprisings." *The Harvard Gazette* (April 9, 2012), https://news.harvard.edu/gazette/story/2012/04/harvards-long-ago-student-risings/.

Issapour, Marjaneh and Keith Sheppard. "Evolution of American Engineering Education." Presented at the 2015 Conference for Industry and Education Collaboration, sponsored by the American Society for Engineering Education, http://www.indiana.edu/~ciec/Proceedings_2015/ETD/ETD315_IssapourSheppard.pdf.

James, Frank A. J. L., ed. *The Development of the Laboratory: Essays on the Place of Experiment in Industrial Civilization*. London: Macmillan, 1989.
James, Mary Ann. "Engineering an Environment for Change: Bigelow, Peirce, and Early Nineteenth-Century Practical Education at Harvard." In *Science at Harvard University: Historical Perspectives*, edited by Clark Elliott and Margaret W. Rossiter. Bethlehem, PA: Lehigh University Press, 1992.
Jenks, Henry F. *The Boston Public Latin School, 1635-1880*. Cambridge, MA: Moses King, 1881.
"John Leverett, 1662/3-1724." In *The Bloomsbury Encyclopedia of American Enlightenment*, edited by Mark Spencer. New York & London: Bloomsbury Academic, 2015, 2:624.
Katula, Richard A. *The Eloquence of Edward Everett: America's Greatest Orator*. New York: Peter Lang, 2010.
Katz, Michael S. *A History of Compulsory Education Laws*. Phi Delta Kappa Fastback Series, no. 75. Bloomington, IN: Phi Delta Kappa Educational Foundation, 1976.
Kennedy, Lawrence, W. *Planning the City Upon a Hill: Boston since 1630*. Amherst, MA: University of Massachusetts Press, 1992.
Kern, Julie. "The Yale Report of 1828: A Synopsis," https://www3.nd.edu/~rbarger/www7/yalerpt.html.
Khrapak, Vyacheslav. "Reflections on the American Lyceum: The Legacy of Josiah Holbrook and the Transcendental Sessions." *Journal of Philosophy & History of Education* 64, no. 1 (2014): 47-62.
Kidder, James. *A View of the State House from the Common*. Boston Public Library's Pictorial Archive, http://www.digitalcommonwealth.org/search/commonwealth:c821gs31f.
Kimball, Bruce A. *The Liberal Arts Tradition: A Documentary History*. Lanham, MD: University Press of America, 2010.
Kingsley, James Luce. *A Sketch of the History of Yale College in Connecticut*. Boston: Perkins, Marvin, & Co., 1835.
Klee, Jeffrey Eugene. "Civic Order on Beacon Hill, 1790-1850." Dissertation submitted in partial fulfillment of the requirements for the degree of Doctor of Philosophy in Art History at the University of Delaware, 2016.
Klein, Christopher. "The Man Who Shipped New England Ice Around the World," http://www.history.com/news/the-man-who-shipped-new-england-ice-around-the-world.
Krieger, Alex, and David Cobb, eds. *Mapping Boston*. Boston: The Muriel G. and Norman B. Leventhal Family Foundation, 1999.
"Lawrence Scientific School," Barnard, Henry, and Absalom Peters, eds. *The American Journal of Education and College Review* 2 (January 1856): 217-224.

Lenney, John J. *Caste System in the American Army: A Study of the Corps of Engineers and Their West Point System*. New York: Greenberg Publishing, 1949.

Li-Marcus, Moying. *Beacon Hill: The Life and Times of a Neighborhood*. Boston: Northeastern University Press, 2002.

Liebig, Justus von and H. Kopp. *Annual Report of the Progress of Chemistry, and the Allied Sciences, Physics, Mineralogy, and Geology*. vol. I, 1847-1848. London: Taylor, Walton, & Maberly, 1849.

Lingwall, Jeff. *An Economic History of Compulsory Attendance and Child Labor Laws in the United States, 1810-1926*. Dissertation submitted in partial fulfillment of the requirements for the degree of Doctor of Philosophy in Economics and Public Policy at Carnegie Mellon University, 2014.

List of Students of the Lawrence Scientific School, 1847-1894. Cambridge, MA: Harvard University Press, 1898.

Lodge, Henry Cabot. *Early Memories*. New York: Charles Scribner's Sons, 1913.

Logel, Jon Scott. *Designing Gotham: West Point Engineers and the Rise of Modern New York, 1817-1898*. Baton Rouge: Louisiana State University Press, 2016.

Lovell, Margaretta M. *Art in a Season of Revolution: Painters, Artists, and Patrons in Early America*. Philadelphia, PA: University of Pennsylvania Press, 2005.

Lovett, James D'Wolf. *Old Boston Boys and the Games They Played*. Boston: Little, Brown & Co., 1908.

Lowell, John. *Extracts from the Will and Codicil of John Lowell, Jr, Concerning the Lecture Fund and Opinions of Hon. Chas. G. Loring, and Hon. Benj. R. Curtis Upon the Duties of the Trustees of the Boston Athenaeum as Visitors of the Lecture Fund*. Boston: Privately Printed, 1885.

Lurie, Edward. *Louis Agassiz, a Life in Science*. Chicago: University of Chicago Press, 1960.

Lyons, Jonathan. *The Society of Useful Knowledge: How Benjamin Franklin and Friends Brought the Enlightenment to America*. New York: Bloomsbury Press, 2013.

Malone, Dumas. *The Sage of Monticello: Jefferson and his Time, Vol. 6*. Boston: Little, Brown & Co., 1981.

Mason, Matthew. *Apostle of Union: A Political Biography of Edward Everett*. Chapel Hill, NC: University of North Carolina Press, 2016.

Mattingly, Paul H. *The Classless Profession: American Schoolmen in the Nineteenth Century*. New York: New York University Press, 1975.

McCaughey, Robert A. *Josiah Quincy, 1772-1864: The Last Federalist*. Cambridge, MA: Harvard University Press, 1974.

McIntyre, Alex McVoy. *Beacon Hill: A Walking Tour*. Boston: Little, Brown & Co., 1975.

McLachlan, James. *American Boarding Schools: A Historical Study*. New York: Charles Scribner's Sons, 1970.
Memoir of the Boston Athenaeum. With the Act of Incorporation, and Organization of the Institution. Boston: Munroe & Francis, 1807.
Memorial of Edward Everett from the City of Boston. Boston: J. E. Farwell & Co., 1865.
"Memorial of Joseph Lovering, Late President of the Academy." *American Academy of Arts and Sciences*. Cambridge: John Wilson & Son, 1892.
Meyer, William B. "A City (Only Partly) on a Hill: Terrain and Land Use in Pre-Twentieth-Century Boston." In *Remaking Boston: An Environmental History of the City and Its Surroundings*, edited by Anthony N. Penna and Conrad Edick Wright. Pittsburgh: University of Pittsburgh Press, 2009.
Morison, Samuel Eliot, ed. *The Development of Harvard University, 1869-1929*. Cambridge, MA: Harvard University Press, 1930.
_____. *Three Centuries of Harvard, 1636-1936*. Cambridge, MA: Belknap Press of Harvard University, 1965.
Nash, Jr., Nathaniel, and William Leavitt Stoddard, eds. *Official Guide to Harvard University Edited by The Harvard Memorial Society*. Cambridge, MA: Harvard University Press, 1907.
New England's First Fruits: In Respect of the Colledge, and the Proceedings of Learning Therein. Old South Leaflets, 12th Series, No. 51. Boston: Old South Meeting House, 1894.
O'Connor, Thomas H. *Bibles, Brahmins and Bosses: A Short History of Boston*. Boston: Trustees of the Public Library of the City of Boston, 1984.
O'Hara, R. J. "The Yale Report of 1828, Part I & II." *The Collegiate Way: Residential Colleges & the Renewal of University Life*, http://collegiateway.org/reading/yale-report-1828/.
One Hundred Years of the English High School of Boston. Boston: English High School Association, 1924.
Pak, Michael S. "The Yale Report of 1828: A New Reading and New Implications." *History of Education Quarterly* 48, no. 1 (Feb. 2008): 30-57.
Parker, Samuel Chester. *A Textbook in the History of Modern Elementary Education, with Emphasis on School Practice in Relation to Social Conditions*. Boston: Ginn & Co., 1912.
Parsons, Theophilus. *Memoirs of Charles Folsom: From Proceedings of the Massachusetts Historical Society, 1873*. Cambridge, MA: John Wilson & Son, 1873.
Peirce, B. Osgood. "Biographical Memoir of Joseph Lovering, 1813-1892." *National Academy of Sciences Biographical Memoirs* 6 (1909): 329-344.

Peirce, Benjamin. *A History of Harvard University from its Foundation in the Year 1636 to the Period of the American Revolution*. Cambridge, MA: Brown, Shattuck & Co., 1883.

Pfammatter, Ulrich. *The Making of the Modern Architect and Engineer: The Origins and Development of a Scientific and Industrially Oriented Education*. Basel, Switzerland: Birkhäuser, 2000.

Pickering, Edward C. "Department of Physics, Jan. 31, 1877," *Massachusetts Institute of Technology President's Report for the Year Ending Sept. 30, 1876*. Boston: A. A. Kingman, 1877.

———. *Elements of Physical Manipulation*. New York: Hurd & Houghton, 1873-76. Pickering, Mary Orne. *Life of John Pickering*. Boston: John Wilson & Son, 1887.

Potter, Alfred C. *Catalogue of John Harvard's Library*. Cambridge, MA: John Wilson & Son, 1919.

Prescott, Samuel C. *When M.I.T. Was "Boston Tech," 1861-1916*. Cambridge, MA: MIT Press, 1954.

Prescott, William Hickling. *Memoir of the Honorable Abbott Lawrence Prepared for the National Portrait Gallery*. Privately Printed, 1856.

Purrington, Robert D. *Physics in the Nineteenth Century*. New Brunswick, NJ: Rutgers University Press, 1997.

Quincy, Josiah. *The History of Harvard University*. Cambridge, MA: John Owen, 1840.

———. *The History of the Boston Athenaeum, with Biographical Notices of its Deceased Founders*. Cambridge, MA: Metcalf & Co., 1851.

Rawson, Michael. *Eden on the Charles: The Making of Boston*. Cambridge, MA: Harvard University Press, 2010.

Ray, Angela G. *Lyceum and Public Culture in the Nineteenth-Century United States*. East Lansing, MI: Michigan State University Press, 2005.

Reed, Roger G. *Building Victorian Boston: The Architecture of Gridley J. F. Bryant*. Amherst, MA: University of Massachusetts Press, 2007.

Remick, Christian. *A Prospective View of Part of the Commons*. In Margaretta Lovell's *Art in a Season of Revolution: Painters, Artists, and Patrons in Early America*. Philadelphia, PA: University of Pennsylvania Press, 2007.

Reports on the Course of Instruction in Yale College; by a Committee of the Corporation, and the Academical Faculty. New Haven, CT: Yale Corporation, 1828.

Resetarits, C. R. *An Anthology of Nineteenth-Century American Science Writing*. New York & London: Anthem Press, 2013.

Reynolds, Terry S. "The Education of Engineers in America before the Morrill Act of 1862." *History of Education Quarterly* 32, no. 4 (1992): 459-482.

Reznek, Samuel. *Education for a Technological Society: A Sesquicentennial History of Rensselaer Polytechnic Institute.* Troy, NY: Rensselaer Polytechnic Institute, 1968.

Rickoff, Andrew J. *Past and Present of Our Common School Education: Reply to President B. A. Hinsdale, with a Brief Sketch of the History of Elementary Education in America.* Cleveland, OH: Leader Print Co., 1877.

Rogers, William Barton. "A Plan for a Polytechnic School in Boston, 1846." In *Life and Letters of William Barton Rogers, vol. I*, edited by Emma Savage Rogers. Cambridge, MA: Riverside Press, 1896.

———. "An Act to Incorporate the Massachusetts Institute of Technology, and to Grant Aid to Said Institute and to the Boston Society of Natural History, Approved April 10, 1861, Chapter 183." *Acts and Resolves of the General Court relating to the Massachusetts Institute of Technology*, https://libraries.mit.edu/archives/mithistory/pdf/1861%20Charter.pdf

———. *An Elementary Treatise on the Strength of Materials, Being the Substance of the Lectures on that Subject, Delivered in the School of Engineering of the University of Virginia.* Charlottesville, VA: Tompkins and Noel, 1838.

———. *Elements of Mechanical Philosophy, for the Use of the Junior Students of the University of Virginia.* Boston: Thurston, Torry, & Emerson, 1852.

———. *For the Establishment of a School of Arts, Memorial of the Franklin Institute, of the State of Pennsylvania, for the Promotion of the Mechanic Arts, to the Legislature of Pennsylvania.* Philadelphia, PA: The Franklin Institute, J. Crissy Printers, 1837.

———. *Life and Letters of William Barton Rogers.* Edited by Emma Savage Rogers with the assistance of William T. Sedgwick, in two volumes. Boston & New York: Houghton, Mifflin & Co., 1896.

———. *Objects and Plan of an Institute of Technology including a Society of Arts, a Museum of Arts, and a School of Industrial Science, proposed to be established in Boston.* Boston: John Wilson & Son, 1860.

———. *Scope and Plan of the School of Industrial Science of the Massachusetts Institute of Technology*, as Reported by the Committee on Instruction of the Institute, and Adopted by the Government, May 30, 1864.

——— and Henry Darwin Rogers. "On the Structure of the Appalachian Chain, as exemplifying the Laws which have regulated the Elevation of Great Mountain Chains generally." In *A Reprint of Annual Reports and Other Papers, on the Geology of the Virginias*, edited by Emma Savage Rogers. New York: D. Appleton & Co., 1884, 601-642.

Routledge, Robert. *A Popular History of Science.* New York & London: George Routledge & Sons, 1881.

Rudolph, Frederick. *The American College and University: A History.* Athens, GA: University of Georgia Press, 1990.

Saltzman, Martin D. "Benjamin Silliman Jr.'s 1874 Papers: American Contributions to Chemistry." *Bulletin of the History of Chemistry* 36, no. 1 (2011): 22-34.

Samuels, Ernest. *Henry Adams.* Cambridge, MA: Belknap Press of Harvard University, 1995.

Seasholes, Nancy S. *Gaining Ground: A History of Landmaking in Boston.* Cambridge, MA: MIT Press, 2003.

Seybolt, Robert Francis. *Source Studies in American Colonial Education: The Private School.* New York: Arno Press, 1971. Originally published by the University of Illinois, Urbana-Champaign, 1925.

―――――. *The Public Schools of Colonial Boston, 1635-1775.* Cambridge, MA: Harvard University Press, 1935.

Schiff, Judith Ann. "Learning by Doing." *Yale Alumni Magazine* (November 2000), http://archives.yalealumnimagazine.com/issues/00_11/oldyale.html.

Schlesinger, Andrew. *Veritas: Harvard College and the American Experience.* Chicago: Ivan R. Dee, 2005.

Scott, Donald M. "The Popular Lecture and the Creation of a Public in Mid-Nineteenth-Century America." *Journal of American History* 66, no. 4 (March 1980): 791-809.

Shurtleff, Nathaniel Bradstreet. *A Topographical and Historical Description of Boston.* Boston: Printed by request of the City Council, 1871.

Sinclair, Bruce. "Harvard, MIT, and the Ideal Technical Education." In *Science at Harvard University: Historical Perspectives*, edited by Clark Elliott and Margaret W. Rossiter. Bethlehem, PA: Lehigh University Press, 1992.

Sinnott, Joseph A. "History of Geology in Massachusetts." In *The State Geological Surveys: A History*, edited by Arthur A. Socolow. Washington, DC: American Association of State Geologists, 1988.

Skelton, William B. "West Point and Officer Professionalism, 1817-1877." In *West Point: Two Centuries and Beyond*, edited by Lance Betros. Abilene, TX: McWhitney Foundation Press, 2004.

Smith, Elske V. P. "Astronomical Laboratory Exercises." *Annals of the New York Academy of Science* 198, no. 1 (August 1972): 124-131.

Smith, Harriette Knight. *The History of the Lowell Institute.* Boston: Lamson, Wolffe & Co., 1898.

Snow, Caleb H. *A History of Boston, the Metropolis of Massachusetts, from its Origin to the Present Period; with Some Account of the Environs.* Boston: Abel Bowen, 1828.

Sobel, Dava. *The Glass Universe: How the Ladies of the Harvard Observatory Took the Measure of the Universe.* New York: Viking, 2016.

Socolow, Arthur A., ed. *The State Geological Surveys: A History.* Washington, DC: American Association of State Geologists, 1988.

Spanagel, David. *DeWitt Clinton and Amos Eaton: Geology and Power in Early New York.* Baltimore, MD: Johns Hopkins University Press, 2014.

Statement of the Course of Instruction, Terms of Admission, &c. at Harvard University, Cambridge, Massachusetts. Cambridge, MA: University Press, Hilliard & Metcalf, 1823.

Story, Ronald. *Harvard & the Boston Upper Classes: The Forging of an Aristocracy, 1800-1870.* Middletown, CT: Wesleyan University Press, 1980.

Stratton, Julius A. and Loretta H. Mannix. *Mind and Hand: The Birth of MIT.* Cambridge, MA: MIT Press, 2005.

The Thayer School of Civil Engineering at Dartmouth College. Hanover, NH: Dartmouth College, 1931.

Ticknor, George. *Life, Letters, and Journals of George Ticknor in Two Volumes,* edited by George S. Hillard, assisted by Anna Ticknor. Boston: James R. Osgood & Co., 1876.

Todd, John E. *Death in the Palace: A Sermon in Memory of Edward Everett, January 22, 1865.* Boston: Dakin & Metcalf, 1865.

Trowbridge, John. *The New Physics: A Manual of Experimental Study for High Schools and Preparatory Schools for College.* New York: D. Appleton & Co., 1884.

Tuttle, Lucius. *An Introduction to Laboratory Physics.* Philadelphia, PA: Jefferson Laboratory of Physics, 1915.

Tyack, David B. *George Ticknor and the Boston Brahmins.* Cambridge, MA: Harvard University Press, 1967.

Tyler, H. W. *John Daniel Runkle, 1822-1902: A Memorial.* Boston: G. H. Ellis Co., 1902.

The U.S. Army Corps of Engineers: A History. Washington, DC: U.S. Government Printing Office, 2008.

Varg, Paul A. *Edward Everett: The Intellectual in the Turmoil of Politics.* Toronto: Associated University Presses, 1992.

Wahl, William H. *The Franklin Institute of the State of Pennsylvania for the Promotion of the Mechanic Arts: A Sketch of its Organization and History.* Philadelphia, PA: Franklin Institute, 1895.

Walker, Alfred E. "Show Me the Way to Old Yale, Boys." In *Trigintennial Record of the Class of 1876 of the Sheffield Scientific School of Yale College.* New Haven, CT: Tuttle, Morehouse & Taylor, 1908.

Warner, Jr., Sam Bass. "A Brief History of Boston." In *Mapping Boston,* edited by Alex Krieger and David Cobb. Boston: The Muriel G. and Norman B. Leventhal Family Foundation, 1999.

Warren, Charles H. "The Sheffield Scientific School from 1847 to 1947." In *The Centennial of the Sheffield Scientific School*, edited by George R. Baitsell. New Haven, CT: Yale University Press, 1950.

Wechsler, Harold S., Lester F. Goodchild, and Linda Eisenmann, eds. *The History of Higher Education*. Boston: Pearson Custom Publishing, 2007.

Whitehead, John S. *The Separation of College and State: Columbia, Dartmouth, Harvard, and Yale, 1776-1876*. New Haven, CT: Yale University Press, 1973.

Whitehill, Walter Muir. *Boston: A Topographical History*. Cambridge & London: Belknap Press of Harvard University, 1959, 1968; revised and expanded by Lawrence W. Kennedy, 2000.

Wilder, Joan. K. *Charles William Eliot and American Education Reform, 1909-1926*. Dissertation submitted in partial fulfillment of the requirements for the degree of Doctor of Philosophy in Educational Policy Studies at the University of Wisconsin, 1970.

Williams, Mari E. W. "Astronomical Observatories as Practical Space: The Case of Pulkowa." In *The Development of the Laboratory: Essays on the Place of Experiment in Industrial Civilization*, edited by Frank A. J. L. James. London: Macmillan, 1989.

Wilson, Leonard G. "Benjamin Silliman: A Biographical Sketch." In *Benjamin Silliman and His Circle: Studies on the Influence of Benjamin Silliman on Science in America*, edited by Leonard G. Wilson. New York: Science History Publications, 1979.

Wilson, Thomas L. V., *The Aristocracy in Boston: Who They Were and What They Were: Being a History of the Business and Business Men in Boston*. Boston: Private Printing, 1848.

Wright, Tom F., ed. *The Cosmopolitan Lyceum: Lecture Culture and the Globe in Nineteenth-Century America*. Amherst, MA: University of Massachusetts Press, 2013.

Wylie, Francis E. *M.I.T. in Perspective: A Pictorial History of the Massachusetts Institute of Technology*. Boston: Little, Brown & Co., 1975.

Name Index

A

Adams, Henry 36
Adams, John 56
Adams, John Quincy 4, 33, 117, 121
Agassiz, Louis 148, 171–172, 183–184, 196, 197–198, 199, 204, 232, 250, 256–257, 261
Andrew, John Albion 178, 180–181, 183, 184
Arnold, Benedict 74
Atkinson, William 192, 259, 260

B

Bache, Alexander 144, 197, 232, 261, 262
Baker, Amos Prescott 39
Banks, Nathaniel 171–172
Barton, Benjamin Smith 96–97, 152
Beck, Charles 118–119, 120
Benjamin, Asher 236
Bigelow, Erastus 255, 257
Bigelow, Jacob 71, 125–126, 135, 258
Bigelow, William 51
Blackstone, William 8
Bôcher, Ferdinand 191
Bond, William 144, 145
Boutwell, George 171, 172, 175, 180
Bowditch, Charles 2, 235
Bowditch, Nathaniel 115, 136
Brattle, William 109
Brooks, Phillips 54
Bryant, Gridley 134, 211

Bulfinch, Charles 11, 16, 20, 21, 51, 136, 236, 260

C

Campbell, Colen 237
Carlton, W. T. 191
Carter, James 30–31
Chandler, Abiel 131, 133, 231
Channing, William Ellery 71, 136
Charles Chauncy 38
Chase, John 258
Cheever, Ezekiel 48–49, 50, 54, 55, 57, 238–239
Clark, Jonas Gilman 234
Claxton, Timothy 69–70, 227, 230, 241
Cook, George 80–81
Cooke, Josiah 147–148, 191, 193–194, 195–196, 197, 231, 263
Coolidge, Cornelius 13
Cooper, Peter 133, 232
Copley, John Singleton 10, 18
Cornell, Ezra 233
Cotting, Uriah 18, 171
Cotton, John 44–45, 246
Cushing, Thomas 38

D

Dalton, Charles 255, 258
Dana, James Dwight 103, 261
Darwin, Charles 196, 250, 256–257
Dawes, Thomas 138
Day, Jeremiah 88–92, 96, 100, 103

---

*Please note that the name index is not comprehensive. It omits names of individuals appearing for the first—and often, only—time in a direct quotation, a serial listing, or a parenthetical aside (such as the progenitors of an individual, i.e., parents or grandparents, as illustrious as they may be). I have also been selective in adding names from the chronology and chapter notes, as not every name contributes directly to the theme of the book. Finally, although the first chapter reviews the entire Pickering family tree, I have listed only Edward Pickering and his younger brother, William Henry Pickering, and their grandfather, John Pickering, as other relatives discussed do not play a major role in furthering the main thesis of this book.

Dearborn, Edmund 39
Dearborn, Henry 135
Derby, Richard 21
Dillaway, Charles 52
Dimmock, William 53
Dixwell, Epes Sargent 35–37, 52–53
Drexel, Anthony 234
Dummer, William 111
Dunster, Henry 30, 55, 108
Dwight, Timothy 92–96, 101

E

Eaton, Amos 78–81, 224, 227, 258
Eaton, Nathaniel 107
Eliot, Charles William 58, 59, 141, 142, 149, 192–209, 211, 225, 226, 233, 247, 263, 265
Eliot, Samuel A. 121, 131
Ellery, Harrison 2, 235
Emerson, George 255
Eustis, Henry 141, 196
Everett, Edward 71, 120–125, 127, 130–131, 133

F

Farrar, John 144, 146–147, 249
Franklin, Benjamin 57, 88, 111, 243, 252, 258
Fraser, James 65

G

Gardiner, Robert Hallowell 229
Gardner, Francis 40–41, 53–55, 59, 141, 209
Gibbs, George 96, 198
Gibbs, Wolcott 198–199, 202, 261, 262
Gilman, Arthur 211
Gilman, Daniel Coit 182, 245
Gorges, Robert 8
Gould, Benjamin 51–52, 261
Gray, Asa 197, 257
Gray, Francis Calley 171
Greene, Benjamin 80–82, 227

H

Hague, James 192, 260
Hammond, Daniel 7
Hancock, John 9, 10, 19, 111, 239
Hancock, Thomas 10

Hare, Robert 198
Harvard, John 108, 110, 139
Henck, John 192, 209
Henry, Joseph 261
Hill, Thomas 197, 198, 207, 262
Holbrook, Josiah xii, 66–71, 79, 81, 99, 154, 227, 230, 241, 252
Hollis, Thomas (elder) 111, 138
Hollis, Thomas (younger) 138–139
Holmes, Oliver Wendell 16, 166, 254
Holworthy, Matthew 139
Hopkins, Johns 233
Horsford, Ebenezer Norton 127, 129, 131, 132, 141, 142, 195–196, 197, 249, 262
Hunt, Samuel 51
Hutchinson, Thomas 48

J

Jefferson, Thomas 13, 18, 74, 75, 126, 153, 229, 230, 241
Johnson, Samuel 88
Joy, John 18

K

Keating, William 229
King, Rufus 5
Kingsley, James 88–89
Kirkland, John Thornton 113–115, 116, 121, 140
Kneeland, Samuel 172

L

Lawrence, Abbott 59, 61, 130–133, 171, 231
Lawrence, Amos 133
Leverett, Frederic 52
Leverett, John 109–111, 246
Liebig, Justus Von 129, 198
Lincoln, Abraham 14, 135, 182
Lodge, Henry Cabot 37, 52
Loring, Charles 36–42
Lovell, James 51
Lovell, John 50–51, 56
Lovering, Joseph 144–149, 197
Lowell, John Amory 72, 162–165, 201, 258
Lowell, Jr., John 72, 230
Lyman, George Williams 19

NAME INDEX

M

Maclean, John 94
Mahan, Dennis Hart 242–243
Mann, Horace xii, 31–33, 53, 62, 121, 238
Mason, Jonathan 10, 11, 13, 17, 19, 20, 236
Mather, Cotton 55, 56, 57, 239
Mather, Increase 109, 112
Maude, Daniel 46, 47
McCook, James 233
McIntyre, Samuel 21
McRee, William 75
Merrick, Samuel 229
Morrill, Justin Smith 181–182

N

Norton, John Pitkin 100–101, 103, 105, 244–245
Norton, William 102
Nott, Eliphalet 97, 205, 230, 244

O

Otis, Harrison Gray 10, 11, 13, 15, 16–17, 20, 21, 136, 236

P

Parris, Alexander 17, 19
Partridge, Alden 76, 79–80
Peabody, Andrew 145–146, 207–208
Pearson, Eliphalet 112–113
Peirce, Benjamin 118, 119–120, 125, 126, 130, 131, 144, 145, 146–147, 149, 179, 191, 195, 197, 198, 199, 249, 261
Phelps, Franklin 40
Phillips, John 18
Phillips, Samuel 113
Phillips, Wendell 18
Pickering, Edward Charles x–xii, 1–3, 6, 8, 10–12, 16, 17, 18, 19, 21, 22–23, 25, 33, 35, 37–38, 39, 40–41, 43–44, 54–55, 58–59, 61, 105, 133–134, 136–139, 141–144, 149, 209, 211, 213–215, 217–219, 221–222, 224–227, 232, 249, 251, 264
Pickering, John 2–3, 5–6, 38, 71, 235, 241, 281

Pickering, William Henry 8, 59, 281
Pormort, Philemon 46, 47
Porter, John Addison 103
Pratt, Charles 234
Preston, Jonathan 211, 259
Preston, William 211, 259
Putnam, Samuel 5

Q

Quincy, Josiah 116–118, 119–120, 122, 247

R

Reed, Roger 39
Rensselaer, Stephen Van 22, 78, 230
Revere, Paul 62
Rogers, Henry Darwin 151, 153, 154, 155–158, 161, 162–163, 166, 167, 169–170, 173, 253
Rogers, James Blythe 151, 152, 154, 155, 156, 167, 253
Rogers, Patrick Kerr 151, 152, 153–154, 155
Rogers, Robert Empie 151, 153, 154, 155, 156, 166, 167, 170, 253, 254
Rogers, William Barton 149, 151, 153, 154–167, 169, 183, 185, 187–188, 189, 191, 200, 204, 209, 214, 218, 224, 227, 232, 251, 253, 254, 255, 256, 257, 258, 260, 261, 264, 266
Ross, Matthias 255, 257
Rotch, Benjamin 141
Runkle, John 179, 188–189, 191, 208, 214–215, 217, 218, 222, 257, 264, 266
Rush, Benjamin 86–87

S

Samuels, Ernest 36
Sears, David 17
Sedgwick, William T. 251
Shaw, Robert Gould 18
Sheffield, Joseph E. 103–104, 232
Shippen, Jr. William 152
Silliman, Jr., Benjamin 98, 99–101, 102–103, 105, 244

Silliman, Sr., Benjamin 66, 91–97, 98–100, 102–103, 104, 105, 120, 144, 148, 198, 229, 244
Smith, William 88, 252
Smithson, James 231
Sparks, Jared 136, 193
Sprague, John 29
Stanford, Jane Lathrop 234
Stanford, Leland 234
Stiles, Ezra 93
Storer, Francis (Frank) 179, 191, 194, 198, 200–201, 207, 208–209, 227, 260, 262, 263, 266
Sullivan, Thomas Russell 39
Swallow, Ellen 259
Swan, Hepzibah 13, 20

T

Tappan, Henry 231
Thayer, Gideon 38
Thayer, John 22, 264
Thayer, Nathaniel 22, 264
Thayer, Sylvanus 74–75, 76–78, 80, 227, 229, 233
Thompson, Benjamin 125, 229, 248
Thorndike, Israel 18
Throop, Amos 234
Ticknor, George 114–115, 121, 193, 246
Tilgman, Edward 5
Tompson, Benjamin 48
Treadwell, Daniel 126–127, 130
Trowbridge, John 251
Tudor, Frederic 19, 237

U

Upjohn, Richard 22

W

Wadsworth, Alexander 135
Wadsworth, Benjamin 136
Walker, Francis Amasa 264
Ware, William 192
Warren, John C. 231
Watson, William 191
Waud, William 173
Wayland, Francis 206
Webb, Thomas 258
Webber, Samuel 112, 114

Webster, Daniel 71
Webster, John White 129
Weston, Edward 233
White, Andrew 233
White, Joseph 180
Wilder, Marshall 258
Willard, Joseph 112
Willard, Samuel 109
Williams, Nathaniel 50, 55, 239
Winthrop, John 8, 18
Winthrop, Thomas 18
Wistar, Caspar 94, 152
Wood, William 63
Woodbridge, John 46, 47
Woodhouse, James 94, 152
Woodmansey, Robert 46, 47–48
Woolsey, Dwight 102–103
Wyman, Jeffries 197

About the Author

W. Nikola-Lisa is professor emeritus at National-Louis University in Chicago, Illinois, where he taught in the Graduate School of Education. He is the author of numerous books for a variety of age levels. His books include the award-winning *How We Are Smart*, an exploration of Howard Gardner's theory of multiple intelligences, *Magic in the Margins*, a story about bookmaking in the Middle Ages, and *The Men Who Made the Yankees*, an homage to one of the greatest baseball franchises in history. For more information visit http://www.nikolabooks.com.

ALSO AVAILABLE

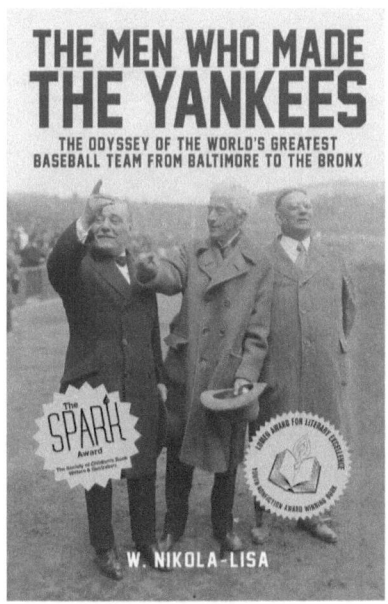

If you liked this book by W. Nikola-Lisa, you also might enjoy *The Men Who Made the Yankees: The Odyssey of the World's Greatest Baseball Team from Baltimore to the Bronx*, a book about politics, money, and ambition, and the role they played in establishing one of major league baseball's greatest franchises. Available in hardcover, paperback, and e-book at all major online booksellers.

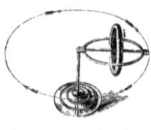

Gyroscope Books

www.ingramcontent.com/pod-product-compliance
Lightning Source LLC
Chambersburg PA
CBHW020734020526
44118CB00033B/596